Isaac Newton

Lives and Legacies

Larzer Ziff
Mark Twain

David S. Reynolds
Walt Whitman

Edwin S. Gaustad
Roger Williams

Gale E. Christianson
Isaac Newton

Isaac Newton

Gale E. Christianson

OXFORD
UNIVERSITY PRESS
2005

OXFORD
UNIVERSITY PRESS

Oxford University Press, Inc., publishes works that further
Oxford University's objective of excellence
in research, scholarship, and education.

Oxford New York
Auckland Cape Town Dar es Salaam Hong Kong Karachi
Kuala Lumpur Madrid Melbourne Mexico City Nairobi
New Delhi Shanghai Taipei Toronto

With offices in
Argentina Austria Brazil Chile Czech Republic France Greece
Guatemala Hungary Italy Japan Poland Portugal Singapore
South Korea Switzerland Thailand Turkey Ukraine Vietnam

Published by Oxford University Press, Inc.
198 Madison Avenue, New York, New York, 10016
www.oup.com

Library of Congress Cataloging-in-Publication Data
Christianson, Gale E.
Isaac Newton / Gale E. Christianson.— 1st ed.
p. cm.
Includes bibliographical references and index.
ISBN-13: 978-0-19-530070-3
ISBN-10: 0-19-530070-X
1. Newton, Isaac, Sir, 1642–1727.
2. Physicists—Great Britain—Biography.
I. Title.
QC16.N7C51 2005
530'.092—dc22
2005009600

1 3 5 7 9 8 6 4 2
Printed in the United States of America
on acid-free paper

To Miss Rhonda

CONTENTS

Doing easily what others find difficult is talent;
doing what is impossible for talent is genius.

Henri-Frédéric Amiel

Preface

At once mortal and immortal, Isaac Newton was both a living, breathing man and a legend that forever changed the world. Yet so reclusive was he at his most creative that his biographers have succeeded in viewing him only through a glass darkly. During a rare moment of introspection, Newton once compared himself to a boy playing on a seashore, casting about for a few beautiful pebbles that he likened to his greatest discoveries. It is worth noting, however, that even this simple metaphor is misleading. While Newton's home was an island, he never set eyes on the open sea beyond the fens until well after he had grown to manhood. Neither, so far as is known, did he seriously entertain the idea of taking in the wonders of the Continent only a short sail away, or of communing with Europe's foremost scientific minds.

Newton was a loner pure and simple, secure in the knowledge that he was without peers when it came to almost all matters cerebral. The only person who knew him well was his enigmatic chamber fellow John Wickins, who lived with him when he was formulating the calculus, conducting his brilliantly counterintuitive prism experiments, and undergoing his earliest intimations of gravity. When the two men finally parted company after sharing the same lodgings at Cambridge University for some twenty years, Wickins married and moved away to become a successful country parson, leaving behind neither letters nor memoirs regarding his supremely gifted friend.

Perhaps Wickins's silence was at least partly due to the fact that he understood so little of what Newton was about, although he often took the part of Newton's copyist. This was no less true of Humphrey Newton, a distant relative hired by Sir Isaac to stoke the fires of his alchemical furnace during what Humphrey called "spring & fall of the Leaf." Humphrey described his employer as a man possessed, one who slept next to the fire through the night without bothering to remove so much as a single article of clothing. As for what Newton was searching for, Humphrey, like the sorcerer's apprentice, hadn't a clue. "His Pains, his Diligence at those set Times," Humphrey wrote, "made me think he aimed at something beyond the Reach of human Art & Industry."

So it went for thirty years and more until Newton resigned his professorship and set out for the greener pastures of London. There he took charge of the royal mint, where he transformed the coining of England's silver and gold while vengefully running counterfeiters to ground for a date with the hangman's noose. And it was in London, while Newton was presiding over the Royal Society like some grand potentate, that he went to war against the German polymath Gottfried Wilhelm von Leibniz and England's astronomer royal, John Flamsteed.

While I can claim no personal knowledge of Newton, I have spent much of the past twenty-five years reading, thinking, and writing about him. This short biography is the partial distillate of that intractable—indeed sometimes maddening—quest for understanding, and is aimed at the casual reader who wishes to know something more of Newton beyond his "discovery" of universal gravitation. Still, after all that time, I, too, see him through a glass darkly. As one Newton scholar remarked, "It would take another Newton to understand Newton." I am anything but that person, yet neither were all the others who have accepted the challenge of putting his remarkable life on paper. What sustained me was the privilege of standing over Newton's shoulder, a silent witness to the creative process as it unfolded in all its majesty. Think of it: a lone human being bent low over a desk, supplied with nothing more than a quill pen, a pot of homemade ink, and countless sheets of blank paper, calculating precisely how the cosmos goes.

When an awed French mathematician wrote to an English friend and fellow member of the Royal Society, he wanted to know if Sir Isaac Newton eats and drinks and sleeps like other men. Of course, his friend replied, "He assumes nothing and puts himself on a level with all mankind." It was an opinion with which the brilliant astronomer and comet seeker Edmond Halley profoundly disagreed: "Nearer to the gods," Halley declared, "no mortal may approach," lifting Newton to the stars.

Isaac Newton

One

"To Play Philosophically"

At a little before ten o'clock the drum roll began, first distant and then ever louder as the thousands gathering to see the king die wound their way through London's choked streets to the palace of Whitehall. It was January 30, 1649. The weather was exceedingly cold, the sky gray with scattered patches of sunshine, the frost so sharp that great chunks of ice clogged the arches of London Bridge spanning the river Thames.

The previous day fifty-nine judges had signed the death warrant of King Charles I, many of them badgered and bullied by the rebel leader Oliver Cromwell, who was about to assume the title of Lord Protector and rule as a king in everything but name. It was Cromwell and his Roundheads, who cut their hair short in defiance of the wigged nobility and practiced the Puritan religion, who most wanted the king's head. They had fought Charles and his supporters, many of whom were sympathetic to the outlawed Catholic religion, ever since 1642, when civil war erupted in the north of England and spread across the land like a great plague. Now the leader of the defeated forces had to pay.

Fearing that the cold would cause him to shiver and give his enemies the false impression that he was afraid, Charles donned two sets of underclothes before putting on his best suit. Flanked by the Bishop of London and his personal attendant, Herbert, the king stepped out of the palace of St. James and headed through the surrounding park toward the scaffold at Whitehall Palace, whose construction the night before had disturbed his sleep. He walked so fast between the double line of soldiers guarding the way that his escorts could barely keep pace. Once at Whitehall he ate a little bread and drank some wine.

At two o'clock there was a break in the clouds. The crowd, kept at a distance by several rows of soldiers, could just make out figures passing rapidly behind the windows of the banquet hall. Suddenly the king emerged by stepping through a long window and onto the scaffold. He knelt down and laid his head upon the wooden block. The masked executioner tucked the long royal hair beneath a cap, then stepped back while the prisoner spent his final moments in prayer. When he was finished, the king stretched out his arms to signal that he was ready to meet his God, and the ax descended in a swift arc. The assistant executioner lifted the severed head for all to see and exclaimed, "Behold, the head of a traitor!"[1]

Instead of the expected cheer, a long, collective moan issued from the crowd. Fearing trouble, Cromwell ordered the streets cleared and his mounted soldiers sprang into action. The spectators panicked and many were trampled in the frenzy, creating even greater fear of the army and the austere leader of the new government.

Woolsthorpe Manor, a gray stone house situated in England's county of Lincoln, was a lonely place seven years before the king's death as civil war spread and Christmas drew near. Its only inhabitants were a young widow named Hannah Newton and her servant. Hannah's husband Isaac, a yeoman, or prosperous farmer, had died in October at the age of thirty-six and was buried in the churchyard of the village of Colsterworth just across the river Witham, which was visible from the upstairs bedrooms. Isaac and Hannah had been married for only five months before he had taken ill. When he failed to get better, a lawyer was summoned and a will drawn up. Most of his possessions, including one hundred acres of land, the manor house, livestock, grain, and household furnishings, were left to Hannah. Like his father before him, Isaac Newton was unlettered, so he signed the document with the traditional "X."[2] No mention was made of the unborn child Hannah was carrying.

Hannah went into labor sometime on December 24, 1642. It was the time of the full moon, and the baby was born an hour or two after midnight on Christmas morning. He was so frail that two women who attended Hannah were sent to obtain some medicine from a neighbor. Instead of hurrying, they sat down to rest on the way, thinking the child would already be dead. In the years to come Hannah would tell her son that he was so tiny when he was born that he would have fit into a quart pot. Moreover, he was too weak to hold his head upright for feeding and breathing. A special collar was fitted to his small neck to support his head; for a long time he remained much smaller than boys his own age and even after maturity, he was somewhat shorter than the average Englishman of his day.

One week later, mother and child were driven to the family church at Colsterworth, where relatives and friends gathered to see the infant baptized. The pastor dutifully recorded the event in the parish register: "Isaac son of Isaac and Hannah Newton baptized Jan. 1, [1643]."[3]

Isaac was still a toddler when his mother caught the eye of an elderly widower named Barnabas Smith, rector of North Witham, a hamlet little more than a mile southeast of Woolsthorpe. Smith held bachelor's and master's degrees from Oxford University and was very wealthy, thanks to a substantial inheritance from his father. Before Hannah would accept the clergyman's marriage proposal, she insisted that he make a gift of some property to her son and remodel Woolsthorpe Manor, which Isaac was expected to occupy one day as a gentleman farmer. Smith agreed and the marriage took place in January 1646, a month after Isaac's third birthday.

For reasons unknown, it was decided that Isaac should remain at Woolsthorpe in the care of his maternal grandparents, the Ayscoughs. The devastated child, who had never set eyes on his father, was suddenly parted from his mother. As he grew older, he discovered how agonizingly close she remained; by climbing a tree he could view the steeple of North Witham's church in the distance. There was Hannah, but there too was the mysterious stranger who had "stolen" her away. The loss ate at him like an emotional cancer. As a teenager, while undergoing a period of intense religious awareness, Isaac compiled a list of "sins" he had committed over the years. Most were harmless, but number thirteen is quite revealing: "Threatening my father and mother Smith to burn them and the house over them." The next entry is equally disturbing: "Wishing death and hoping it to some."[4] The abandoned son not only possessed a volatile temper, but he nursed grudges and would wait years, if need be, to gain revenge on those he believed had wronged him.

Besides his absent parents, the unstated "some" for whom Isaac wished death may have included his two half sisters and his half brother, all fathered by Smith while in his late sixties. Mary, Hannah, and Benjamin were additional rivals for his mother's attention. When Smith finally died in 1653, Hannah, now a very wealthy widow, returned to Woolsthorpe with the three children in tow. By this time eleven-year-old Isaac had long since learned to insulate himself from human contact by withdrawing into the hidden recesses of his mind.

Hannah was determined that Isaac, unlike his father, should grow up literate. During her absence, he was enrolled in a village school to which he walked back and forth each day. Then, a year after Hannah returned to Woolsthorpe, the twelve-year-old youth entered King's School at Grantham, a market town some seven miles away.

Latin and Greek were the languages of instruction, and during this period the foundations of young Newton's classical education were laid. Bible studies were also an important part of the curriculum, and Isaac became familiar with Hebrew script. Though grammar and literature were the main focus, students also received a limited amount of instruction in arithmetic. Isaac may have been introduced to geometry during this period as well, but, if so, there is no record of it in his surviving notebooks.

Arrangements were made for the youth to board in the home of William Clarke, the local apothecary, whose wife was a close friend of Isaac's mother. The Clarkes lived in the middle of Grantham on High Street and by all accounts gave their young boarder as much freedom as good judgment would allow. According to Dr. William Stukeley, a Grantham native whom Newton befriended years later when the two were living in London, "Everyone that knew Sir Isaac recounts the pregnancy of his parts when a boy, his strange inventions and extraordinary inclination for mechanics." Instead of playing with the other boys after school, he "busied himself in making knicknacks and wood models in many kinds: for which purpose he had got little saws, hatchets, hammers and a whole shop of tools, which he would use with great dexterity."[5]

Isaac visited the construction site of a windmill and was inspired to build a working model of his own. When it was finished, a mouse he called the "miller" was placed inside and supplied with a ration of corn. In reaching for the food, the miller set a wheel in motion, which in turn propelled the mill.

The lad was equally fascinated by kites, which he fashioned from paper in various shapes to determine the designs best suited for sustained flight. He also made lanterns from crumpled paper and lit them with candles while walking to school on dark winter mornings. These he sometimes tied to the tails of his kites at night, frightening the country people who mistook them for passing comets.

Smaller and physically weaker than most of his schoolmates, Isaac attempted to teach them "to play philosophically," in Stukeley's words.[6] When the Lord Protector Oliver Cromwell died in September 1658, a great storm swept over England, giving rise to the superstition that it was the devil riding the whirlwind to claim his lost soul. Taking advantage of a rare opportunity, Isaac entered into a competition with several of the more athletic youths to see who could jump the farthest. By carefully timing the gusts of wind, he outleaped the other boys, much to their surprise and embarrassment. Many years later, Newton remarked to a relative that this was one of his first experiments.

This early interest in the force of moving air was matched by a deep fascination with time. Even before Isaac entered King's School, he had undertaken a study of the sun's motion, tracking its path across the yard, walls, and slate roof of Woolsthorpe

Manor. Using a sharp instrument, he carved two sundials next to a window on the building's south face. He did the same on the side of the Clarke house in Grantham, marking the hours and half hours with pegs. According to Stukeley, "Anybody knew what o'clock it was by Isaac's dial, as they ordinarily called it."[7] Later in life, it was said of Newton that he could tell time by glimpsing the positions of shadows on his study walls, and that these observations were as accurate as the dials of his pocket watch.

The most frequently mentioned of his many mechanical models was his water clock, constructed from a wooden box given to him by William Clarke's brother-in-law. It stood four feet high and had a dial at the top with the numbers of the hours. The device was driven by a piece of wood, which alternately rose and fell according to the rhythmic dripping of water. It was kept in the room where he lived and was occasionally checked by the Clarkes to see what hour it was. Like the youthful Leonardo da Vinci and Benjamin Franklin, Isaac Newton was not simply an aimless childhood tinkerer but a tinkerer playing with ideas and mechanisms.

Isaac lived in the Clarkes' attic, where he covered the walls with charcoal sketches he drew from prints as well as from life. These included the beheaded King Charles I, who had died when Newton was not yet seven; the metaphysical poet and clergyman John Donne; and the headmaster of King's School, Henry Stokes. Birds, beasts, ships, and mathematical schemes filled the remaining space, together with verses attributed to the youth himself, an interest that would soon pass.

There is good reason to believe that Isaac also spent many hours in the company of his landlord, while the apothecary mixed medicines for Grantham's citizens. One of many books that captured Isaac's attention was *The Mysteries of Nature and Art* by John

Bate, the third edition of which was published in 1654, when Isaac was eleven. Bate's recipes for mixing paints and treating common ailments found their way into the boy's early notebooks, as did information on diseases and their cures copied from Francis Gregory's catalog titled *Nomenclatura*. Among the least appetizing is a recipe for an abscess: "Drink twice or thrice a day a small portion of mint and wormwood and 300 millipedes well beaten (with their heads pulled off) suspended in 4 gallons of ale in fermentation."[8]

In light of his interest in books and all things mechanical, one would have expected Isaac to do well in his studies, but that was not the case. Henry Stokes put him in the lowest form, where he ranked second to last out of more than eighty students. Newton himself later admitted that he "continued very negligent" in his studies at King's School.

As the story goes, events took a radical turn one morning as Isaac was walking to class. The boy ranked just above him kicked him in the stomach, which provoked thoughts of revenge. When school was over for the day, Isaac challenged his classmate to a fight, and they went into the nearby churchyard to settle their differences. Though smaller than his opponent, Isaac fought with greater determination, beating his attacker until he capitulated by declaring that he would fight no more. He then rubbed the defeated boy's face against the church wall for good measure. Still this was not enough. Isaac began applying himself to his studies and soon rose to become the top student in the school.

The Grantham years were among the happiest in Newton's life, but Hannah had decided that they must come to an end. Her son was about fifteen when she informed Master Stokes that Isaac must return to Woolsthorpe to begin learning the ways of a respected landowner.

Stokes was appalled. Isaac had convinced the headmaster that his talents were too great to be wasted in the isolated countryside of Lincolnshire. Nor could Isaac see himself as one of the simple folk who were his people, for he had gradually ceased to share their lives. To return to the family farm would mean a life of mental boredom that could easily lead to frustration and bitterness, if not abject depression.

Hannah was adamant, thus setting the stage for a conflict of wills. Isaac returned home and later recorded some of his acts of defiance in the list of his sins compiled a few years later. He was "peevish with [his] mother" and refused to go into the fields at her command. He allowed the sheep to stray onto the property of a neighbor, causing damages that an embarrassed Hannah had to pay for in court. Stukeley learned that "his chief delight was to sit under a tree, with a book in his hands, or to busy himself with his knife in cutting wood for models of something or other that struck his fancy, or he would go to a running stream, and make little millwheels to put into the water."[9] Not only did the sheep stray and the wheat go unplanted, but the adolescent often forgot to return home for meals, a character trait that would resurface in the adult Newton.

Frustrated and annoyed, Hannah sought to change Isaac's behavior by placing him under the supervision of an elderly servant. Together they went to Grantham on Saturday mornings to purchase the needed supplies for the farm and to sell the crops raised in the Woolsthorpe fields. But as soon as they arrived, Isaac would head for his old room above the apothecary shop, where he passed the day reading books stored there by his former landlord's brother Joseph Clarke, an assistant teacher at King's School. On other occasions he simply picked out a comfortable spot under a roadside hedge where he read until the servant collected him on the way home. "No doubt," Stukeley observed, "the man [complained] of this to his mother."[10] No doubt indeed!

Clearly Isaac was living among the creations of his own mind. In many ways they were more real to him than anything he encountered in the external world. Henry Stokes, who was keeping close track of his star pupil, finally decided to intercede on Isaac's behalf. He visited Hannah at Woolsthorpe and told her what a great loss it was to bury such a promising intellect on the family farm. "The only way whereby he could preserve or raise his fortune must be by fitting him for the university."[11] The headmaster even offered to forego the annual forty shilling fee required of all pupils born more than a mile from Grantham, no small sacrifice for a man of modest means.

Hannah could be almost as difficult to budge as her oldest son, forcing Stokes to argue his case several times. In her defense it must be remembered that her careful plans for Isaac were falling to pieces, and with them a mother's dreams. To a woman who was next to illiterate, land was the only thing that mattered. She finally turned to her trusted brother William Ayscough for advice. To her surprise, William sided with Stokes, agreeing that Isaac should return to school. Her options closed, Hannah reluctantly gave her consent. For better or for worse, her son would go on to the university.

Isaac's return to Grantham was welcomed by many, including Catherine Storer, Clarke's stepdaughter. She and her two brothers, Arthur and Edward, had grown close to Isaac while living under the same roof on High Street. As an old woman in her eighties, Catherine was interviewed regarding her famous playmate, whom she described as "a sober, silent, thinking lad never known to play with the boys at their silly amusements." Even more intriguing is the suggestion of a romantic attachment to Newton: "Being brought up together, tis said that he entertained a love for her, nor does she deny it. But . . . it was incompatible with his fortunes to marry, perhaps his studies too."[12]

What form those studies took during Isaac's final months of preparation before departing for Cambridge is unknown. Because Latin was the language of scholars, Stokes almost certainly grilled his greatest pupil in the classics one last time. Works by the ancient poets Pindar and Ovid survive from Newton's personal library; both are inscribed "1659" and both were standard texts in the schools of the period. He also had the use of Barnabas Smith's library of some two or three hundred volumes, mostly religious in content, which had been given to him by his widowed mother. These he kept on shelves of his own making in his upstairs bedroom at Woolsthorpe, the very room in which he would soon labor over some of his most important scientific discoveries and mathematical formulations.

Finally, in 1661, the moment of departure arrived. Displaying the pride of a father, Stokes asked eighteen-year-old Isaac to come to the front of the classroom. With tears in his eyes, the headmaster made a speech praising the young man and urging the other youths to follow his example. On a window ledge of King's School, with the aid of a penknife, Newton left a simple, permanent record to be gazed upon by later generations of curious admirers: "I. Newton."

Two

"My Greater Friend"

Thomas Fuller, a seventeenth-century scholar, described Cambridge's Trinity College as "the stateliest and most uniform college in Christendom."[1] On the massive exterior of Trinity's Great Gate is a statue of its founder, the legendary King Henry VIII, whose six wives set a record among English monarchs. It was from this prospect that Isaac Newton first gazed on the institution that would be his home for almost forty years. Before him stood the chapel, the master's lodge, and the magnificent dining hall, with its beamed ceilings, minstrels' gallery, and huge portrait of Henry, which took six strong men to carry. Beyond the Tudor Gothic buildings spread "the Backs," the green expanse bordering the Cam River. To someone of Newton's unworldly background it must have seemed miraculous and frightening at the same time.

Newton entered the university as a sizar, one whose financial circumstances were such that he had to wait tables, run errands for the fellows, or senior scholars, and waken his wealthier classmates before dawn for morning chapel. Despite her considerable wealth, it appears that Hannah Newton was not going to make things easy for her wayward son, forcing him to earn his keep. Equally humiliating was the fact that Newton was going on nineteen in the summer of 1661, which meant that he was a year or two older than most of those whom he served, setting him even further apart.

As with all undergraduates, he was assigned a tutor, who was both a teacher and something of a substitute parent. In this case it was Benjamin Pulleyn, an obscure professor of Greek about whom little is known. The young scholar once again found himself immersed in the classics—the philosophy of Plato and Aristotle, the art of eloquent communication called rhetoric, logic, ethics, history, and much more. There were lectures to attend, papers to write, notebooks to keep, including a worn brown leather volume containing the inscription: "Isaac Newton, Trinity College, Cambridge, 1661."[2]

This volume, known to scholars as "the philosophical notebook," begins with notes from the works of Aristotle. The beautiful script has been entered with careful attention to detail, as if in reverence for the most profound mind in ancient history, and perhaps of all time. Then, suddenly, the reader is met by dozens of blank pages before encountering a radically different series of entries. At some point in 1663, during his third year at Trinity, Newton began a new section, which he titled "Certain Philosophical Questions." More important, he penned the following revolutionary sentence at the top of the first page: "I am a friend of Plato, I am a friend of Aristotle, but truth is my greater friend."[3] He made his own index of the various subjects he planned to investigate.

These range from "Air," "Earth," and "Matter" to "Time and Eternity," "Soul," and "Sleep." Some headings contain no further entries, but under others he wrote from a few sentences to many pages, forsaking his elegant script for the more cramped writing of a young man in a hurry. He got much of his information from the works of an emerging group of thinkers who helped shape the modern worldview, men known as "natural philosophers," for the term "scientist" would not be coined until the nineteenth century.

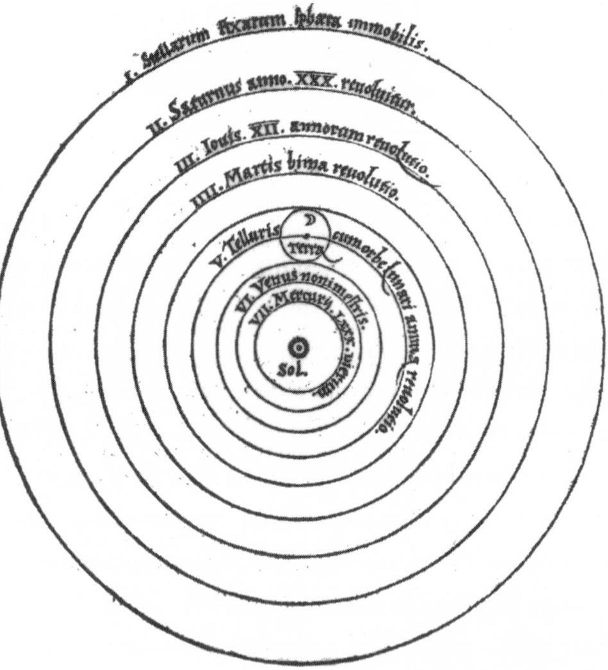

In 1543, the Polish astronomer Nicolaus Copernicus put forth the theory that the Earth and the other planets rotate around the sun, challenging the long-held view that the Earth is the center of the universe. This drawing, from his work *De Revolutionibus,* clearly places the sun at the center of the universe. *Lilly Library, University of Indiana*

Among those whose work he read was Nicolaus Copernicus (1473–1543), the Polish astronomer who, in 1543, had challenged Aristotle and scholarly authority by arguing that the planets, including the Earth, circle the sun. By placing the Earth in motion rather than at rest in the center of the universe, Copernicus had stood a two-thousand-year-old belief on its head.

Others soon stepped in to confirm mathematically and experimentally what Copernicus had presented as theory. Johannes Kepler (1571–1630), the brilliant German astronomer who is now recognized as the first astrophysicist, developed the earliest laws of planetary motion. Kepler's first law states that the shape of each planet's orbit, or path around the sun, is an ellipse, which resembles a gently squeezed balloon. Thus the sun is somewhat off center, and a planet's distance from it varies as it moves about the medium-sized star.

Having shattered the ancient concept that the planets orbit in perfect circles, Kepler next demonstrated in his second law that the speed of a planet is not uniform but varies according to its distance from the sun. When a planet is farthest from the sun, it slows down; conversely, as it approaches the sun, it speeds up. The third law established a relation between the average distance of a planet from the sun and the time it takes to complete a single orbit, or period in the language of astronomers. It would remain for a young Cambridge student to formulate the universal principle by which the planets are held in their orbits, but it was through Kepler's mind and pen that mathematics—elegant, precise, and powerful—entered the heavens.

Newton was also reading the works of the Italian genius Galileo Galilei (1564–1642), who corresponded with Kepler and died in the very year of Newton's birth. Most famous for his use of the

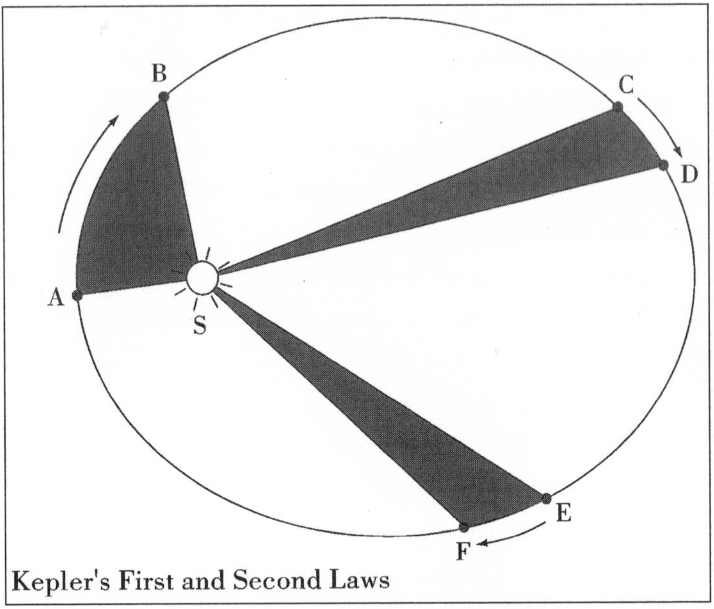

According to the first two laws of planetary motion developed by the German astronomer Johannes Kepler, the planets move around the sun in elliptical orbits and an orbiting planet sweeps over equal areas in equal times. When the planet is far from the sun and moving slower, the pie slice is long and narrow (SCD); when the planet is near the sun and moving faster, the pie slice is short and wide (SAB). *Diagram by Gary Tong*

telescope, Galileo trained the first of his many homemade instruments on the planets, moon, and stars in 1609. What he saw shook the awed natural philosopher to his roots: mountains and craters on the surface of the moon, spots on the sun, the phases of Venus, the moons of Jupiter. And when he looked beyond these objects into the great abyss, he discovered the secret of the Milky Way itself. "The galaxy," he wrote in his revolutionary little treatise *The Starry Messenger,* "is, in fact, nothing but a congeries of innumerable stars grouped together in clusters. Many of them are rather large and quite bright, while the number of smaller ones is quite beyond calculation." As Copernicus had implied by making the Earth a planet, the heavens and the sphere inhabited by humankind are not fundamentally different but are made of the same matter and subject to the same laws.

In 1633 Galileo was condemned by the Roman Catholic Church for teaching what was termed "the Copernican doctrine," which conflicted with the teachings of the Bible. Though he was not tortured or imprisoned, as were others who challenged spiritual authority, he was placed under house arrest for his remaining years, during which he continued to work while going blind from glaucoma.

Newton was no less interested in Galileo's other claim to fame, the experimental method that led to the overthrow of Aristotle's theories. While Kepler was formulating his laws of planetary motion, Galileo developed laws governing the motion of bodies here on Earth. As a professor at the University of Pisa, he found the ancient belief that heavier bodies fall faster than lighter ones to be erroneous. In a series of carefully timed experiments, he was able to show that the acceleration of a falling body is proportional to time and independent of its weight and density. Thus a feather and a cannonball dropped together would reach the ground at the very same time were it not for the resistance of air. This can be demonstrated by employing an airless container, or vacuum chamber.

Moreover, Galileo successfully challenged the time-honored belief that an object at rest is in its natural state. Rather, he postulated, it is natural for an object in motion to continue moving in a straight line, a principle known as inertia. This concept applies equally to a rolling ball on Earth or the circling planet itself. Only when acted upon by some form of resistance do such objects change speed or direction.

Newton was intoxicated by all that he was reading and was preoccupied as much by the behavior of the tiniest objects as by the largest. He embraced those teachings of the ancients that made sense to him, such as the theory of matter set forth by the Greek philosopher Democritus in the fourth century B.C. Democritus held that all things are composed of minute, invisible particles he called atoms. They contain the same matter as larger bodies but differ in size, weight, and shape. According to the philosopher, the constant movement of these whirling particles resulted in the formation of the world and universe as we know it. As Newton wrote in his notebook, "the first matter must be atoms and that matter may be so small as to be indiscernible."[4]

The Trinity notebook further demonstrates that Newton, true to the natural philosopher he was to be, distrusted the senses on which Aristotle and his followers relied. He had already taken a deep interest in the experimental method advocated by Galileo. The senses of different individuals are affected differently by the same phenomena. Newton wrote, "The nature of things is more securely and naturally deduced from their operations on one another than upon the senses; our explanation of both soul and body" must not be colored by either.[5] From then on, when addressing questions to nature, he would embrace the new scientific method and its fundamental steps: gather data, formulate a hypothesis, conduct experiments, validate or reject the hypothesis.

Even as he was studying the works of the new natural philosophers, Newton was practicing science on his own. Fascinated by celestial phenomena, he began tracking comets across the starry sky. He observed the first in December 1664 at 4:30 in the morning, while the rest of the university slept. Another appeared in early April 1665, causing him to wonder how these brilliant objects could move across the firmament with such speed.

Though possessed of an iron will and a physical constitution to match, even Newton had his limits. He later informed John Conduitt, a relative by marriage, that he was so obsessed with nightly observing that he became "much disordered and learned to go to bed betimes."[6] John North, Master of Trinity, observed of Newton that "if he had not wrought with his hands making experiments, he had killed himself with study."[7]

Newton not only courted exhaustion and illness, but risked his eyesight in the interest of science as well. Capturing the sun's image in a looking glass, he observed its rays with his right eye and then turned away. He saw circles of colors, which gradually vanished. Growing bolder, he looked at the sun a second and a third time with both eyes. Suddenly its multicolored imprint overlay every object upon which he gazed, from an open book to a passing cloud. The condition gradually worsened, forcing Newton to shut himself up in his darkened rooms for several days. "I . . . used all means to direct my imagination from the sun. For if I thought upon him I presently saw his picture though I was in the dark."[8] Gradually his sight returned, but even months later, whenever he thought of the sun, its spectrum reappeared "though I lay in bed at midnight with my curtains drawn." When he had sufficiently recovered to take pen in hand, the lingering affliction was duly written up in the philosophical notebook in ten carefully worded steps, a testament to the experimenter's extraordinary self-control.

By 1664 Newton's lifework was pretty much determined. He had joined the intellectual company of those who believed that the universe, like the great clocks adorning the bell towers of medieval cities, is governed by rational mechanical laws and that nature's secrets will be revealed to those employing the scientific method. Yet neither did he think that such precision is the result of blind chance. Behind it all he sensed the presence of intelligent planning and purposeful direction. Under the heading "Of God," he wrote, "Were men and beasts made by fortuitous jumblings of atoms there would be many parts useless in them, here a lump of flesh there a member too much. Some kinds of beasts might have had but one eye some more than two."[9] Above all, this was a reminder to himself that nature's underlying unity is a product of the Divine Mind. Atoms and mechanical laws, no matter how fascinating, are as nothing when compared to the knowledge and wisdom of the Creator.

Newton the mathematician makes no appearance in the philosophical notebook, yet we know from other sources that he was deeply immersed in the study of numbers during this period. According to Abraham de Moivre, a French mathematician and future member of Newton's inner circle, the young Newton visited Sturbridge fair while a Trinity undergraduate and bought a book on astrology out of curiosity. He read until he came to an illustration of the heavens he could not understand for a lack of trigonometry. In turn, he bought a text on trigonometry but could not fully grasp its contents, forcing him to seek out the work of the ancient geometrician Euclid, which seemed absurdly simple. But he soon realized that he had underestimated Euclid's proofs and read the book with greater care a second time. He next graduated to William Oughtred's treatise *Key to Mathematics* and René Descartes's seminal work on analytical geometry, *Discourse on Method,* which, according to de Moivre, "he found very difficult and had to master by degrees."[10] He went back over all that he had read one more time before striking out to claim new territory of his own.

Nowhere is there any mention of a guiding hand, thus leading one to believe that Newton alone was responsible for his mathematical education, as for so much else. Pulleyn, his tutor and a professor of Greek, could have provided at best a measure of encouragement but little, if any, instruction in this field. On at least two occasions in later life, Newton commented on self-education in mathematics and natural philosophy in a manner too suggestive of his own experience to be dismissed as coincidence. He had begun his great quest alone and he would end it the same way, cut off from the interests of the less gifted, locked up in a private world of protective isolation. The British naturalist Charles Darwin knew this world as well. Among Darwin's most rewarding experiences was tramping across the highlands of South America in the silent company of illiterate natives: "The whole of my pleasure was derived from what passed in my mind," he wrote in his journal. And when, after he had grown old and was revered like a god, Newton was asked how he had made his discoveries, he replied, "Truth is the offspring of silence and unbroken meditation."[11]

The one person who seems to have pierced that silence was John Wickins, son of the master of Manchester Grammar School. Wickins told his son Nicholas about his chance encounter with Newton while the two were undergraduates at Trinity. A dejected Wickins had left his rooms to go for a walk one day after an argument with his boisterous roommate. Once outside, he encountered an equally dejected Newton, who was experiencing the very same problem: "They thereupon agreed to shake off their present disorderly companions and chum together, which they did as soon as conveniently they could and so continued as long as my Father stayed at college."[12] This relationship lasted almost twenty years. Unfortunately, John Wickins never recorded his invaluable memories of the most creative period in Newton's life.

Content in their quarters, Newton and Wickins disdained the drinking, gambling, and other antics common among their fellow students. But while Newton confessed to no significant sins of the flesh in his 1662 list of transgressions, he did admit to "setting my heart on money" more than God. This was followed by a further confession that he had suffered "a relapse."

After he celebrated his twenty-first birthday in 1663, the considerable profits from the land deeded to him by the Reverend Smith were his to do with as he pleased. No longer a waiter on tables or an errand runner, Newton the budding natural philosopher became Newton the student moneylender. The philosophical notebook contains a list of his accounts, and judging by the number of clients he served business was thriving. Being the conservative that he was, he only once lent as much as a pound, noting that the sum was "to be paid on Friday."[13] It's unclear how much interest he charged. But since the moneylender has never been a popular figure in society, this practice doubtless only served to set Newton further apart from his fellow students.

In 1664 he passed the required scholarship examinations, ridding himself of the distasteful title of "sizar." Officially declared a "scholar," he was now entitled to receive free meals from his college in addition to a regular stipend. Even more important, he could remain at Trinity to take his master's degree and, if all went well, extend his residence at Cambridge indefinitely by obtaining a fellowship, a plan shared by the steadfast Wickins.

At this point Catherine Storer's little remaining hope of marrying her childhood sweetheart vanished forever, and she accepted the proposal of a Grantham attorney named Francis Bakon. Still Newton remained her friend through the years, visiting her whenever he returned to Lincolnshire, sometimes bearing small gifts.

Less than a year after being made scholar, Newton, together with his fellow candidates for the B.A., took the centuries-old examinations that constituted the last hurdle before gaining a degree. Based mostly on the teachings of ancient and medieval thinkers, the exams placed a premium on oral debate and logic. As a student at Cambridge a generation later, William Stukeley heard that Newton was put in second place, "which is looked upon as disgraceful." Yet, as Stukeley noted, this was hardly out of character: "It seems no strange thing, notwithstanding Sir Isaac's great parts, for we may well suppose him too busy in the solid part of learning to allow much time to be master of words only, or the trifling niceties of logic, which universities still make the chiefest qualification for a degree."[14]

Whatever the truth of the matter, Newton and twenty-five other Trinity men took their degrees in the spring of 1665. While he may have done poorly on the outdated exams, the Trinity notebook stands as proof that truth was indeed his "greater friend," that at Cambridge Isaac Newton had found his genius.

Three

Of Genius, Fire, and Plague

Newton had no sooner graduated from Cambridge than he was forced to leave the university and return to Woolsthorpe, though not of his own choosing. In London, where the city's rapidly increasing population had reached half a million by 1660, people were suddenly taken ill. The disease began with a severe headache and dizziness, followed by trembling in the limbs, swelling under the arms and in the groins, raging fever, and, finally, telltale dark splotches on the skin, which the widely read London cleric Thomas Vincent described as "tokens of near approaching death. Within a few hours they must go down into the dust."[1]

The infamous Black Death, or bubonic plague, had first appeared in Europe in 1347, carried from the Middle East by ship rats in whose fur infected fleas lay hidden. Once a rat succeeded in reaching shore, its parasites were easily transferred to a human host, launching a deadly epidemic. Since there was no known cure for the disease, lord, serf, merchant, and beggar alike fell victim.

In some areas, well over half the population died; commerce and agriculture came to a halt; cities and towns were deserted; wolves reclaimed the silent streets.

Aside from ordering that the ill be completely isolated, physicians could do little but invoke an adage often recited by the ancient Romans whenever they were menaced by disease: "fly quickly, go far, return slowly." Thus, as Samuel Pepys noted in his celebrated diary, the government abandoned London for the relative safety of Oxford, and was soon followed by ordinary citizens who spread across the countryside by the tens of thousands. In September 1665, as the contagion reached its peak, some eight thousand deaths were recorded in the city every week. What was worse, in attempting to save themselves Londoners carried the plague to the eastern counties, and then to the Midlands. In October, Cambridge University's Senate voted to close the institution, but by this time all but a few administrators had fled before the invisible enemy.

Left to his own devices in rural Lincolnshire, Newton began one of the greatest intellectual odysseys in the history of modern science. Thinking back on the two plague years of 1665 and 1666 decades later, he wrote the French scholar Pierre Des Maizeaux, "For in those days I was in the prime of my age of invention and minded mathematics and philosophy more than at any time since."[2]

William Whiston, a future friend and disciple of Newton, wrote that "Sir Isaac, in mathematics, could sometimes see almost by intuition, even without demonstration and when he did propose conjectures in natural philosophy, he almost always knew them to be true at the same time."[3]

Having mastered the works of the most advanced mathematicians of his day, Newton was far from satisfied. Algebra was fine for determining the numerical answer to specific conditions set forth in an equation; geometry served to determine the relationships of points, lines, and angles. But what of more complex questions involving the ever-changing speed of a moving object and the constant alteration of that object's path as its speed changes? Here the mathematician is faced with two quantities that vary continuously, the one with the other, and must determine the rates of variation for any given moment in time.

Others had tackled such numbing problems in the past, and with some measure of success. The French mathematicians René Descartes and Pierre de Fermat (1601–1665) had solved individual problems of this nature, but at the cost of elegance and simplicity. The methods they employed were too cumbersome for general use. Nor had they succeeded in devising a general method that could be applied to all problems of a given type, the stuff of which the great mathematicians dream.

Newton composed his first major mathematical paper at Cambridge in May 1665, shortly before the university closed its doors. He completed a second, more advanced paper at Woolsthorpe in November and three others during the feverish year 1666. His work involved a revolutionary method he termed "fluxions," or quantities subject to constant rates of change. By late 1666 Isaac Newton, who was approaching his twenty-fourth birthday, had become the most advanced mathematician the world had yet known. With his fluxions, or what we now call calculus, a mathematician can solve problems in which quantities and motions are not definite and unchanging but in the process of originating, fluctuating, even disappearing. With these equations Newton could determine the most minute variations in the acceleration of a body as it falls through space, compute the precise arc of a revolving planet, or measure the exact rate in the slowing of a ball as it inches to a stop after being rolled across the ground. In sum, calculus is a most effective tool both for resolving problems concerning the smallest—or what Newton himself termed "infinitesimal"—variations in the rate of motion of a given body and for determining its path through space. Recognizing the need for the utmost precision in our dealings with nature, he had surpassed all others by formulating a mathematical method that deals with nature on its own terms.

Not surprisingly, this newly discovered power thrilled the young man who possessed it. When he visited the nearby home of Humphrey Babington, an older Cambridge scholar who may have helped him get into the university, Newton brought his papers with him. His excitement on solving one mathematical problem was so great that he carried out the division to fifty-five places, and might have continued on had not the long line of zeros literally overflowed the edge of the page. Yet where others would have shouted their victory to the mathematical world, Newton remained curiously silent. Proof to him was sufficient unto itself; the rest of the world did not matter, at least for the time being.

While at Woolsthorpe waiting for Cambridge to reopen, Newton experienced the insight that has since become legend. A year before Newton died in 1727, his friend William Stukeley visited him at his home in Kensington near London. After dining, they went into the garden to drink tea under the shade of some apple trees. "Amidst other discourse," Stukeley wrote, "he told me he was just in the same situation as when formerly the notion of gravitation came into his mind. It was occasioned by the fall of an apple, as he sat in a contemplative mood."[4]

It had occurred to Newton that the power of gravity does not seem to decrease significantly as an object moves away from the center of the Earth. If it remains the same on the highest mountain and atop the loftiest building, "Why not as high as the moon?" he said to himself. If this is true, the moon must be influenced by this mysterious force, "perhaps she is retained in her orbit thereby."[5] What is true of the moon as it orbits Earth must also be true of the planets as they circle the sun. And why not the glimmering stars beyond?

Having read Galileo and Descartes, Newton was well acquainted with the principle of inertia, which holds that once a body is set in motion it will not rest unless acted upon by some external force. He also knew that a body in motion will move in a straight line unless it is diverted somehow. Thus, he conjectured that the natural motion of a ball—or an apple—set in motion would carry it away from Earth, outward into the abyss. But something happens: gravity intervenes, altering the object's course and bringing it to ground.

Instead of throwing a ball, Newton thought of shooting it from a cannon. In this instance it would travel a much greater distance, yet, as when thrown, it would eventually return to Earth. He dreamed of even bigger cannons and larger arcs, and drew sketches to demonstrate how the object would travel farther and farther around the Earth until, finally, it would leave the planet and settle into orbit, like the moon.

Newton was immediately faced with another problem. While the power of gravity does not appear to diminish as one moves away from the Earth's center, this must be an illusion. Were it not so the moon would surely crash into the planet, while the Earth itself, together with Mercury, Venus, Jupiter, Mars, and Saturn, would careen into the sun. Only if gravity decreases at a distance can this celestial balancing act be maintained.

A diagram from Newton's *A Treatise of the System of the World,* published after his death, shows that if a projectile is fired with enough force, it will remain in orbit around the Earth, the same way that the moon is in orbit around the Earth. *Library of Congress*

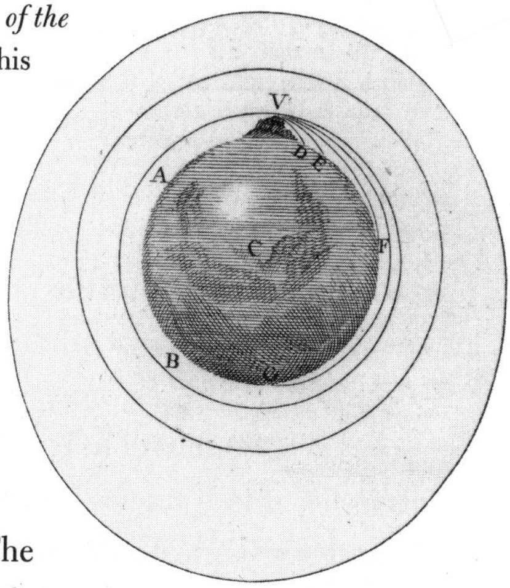

Envisioning the moon as a giant apple, Newton developed the idea of how gravity works. The satellite's tendency to move away from Earth in a straight line is counteracted by the inward pull of gravity, which produces an orbit, much as an object on a string when it is whirled around one's head. The moon is perfectly balanced between the tendency to move outward—or what is termed centrifugal force—and the inward pull of Earth.

At this point Newton immersed himself in mathematical calculations, most of which are lost, in an attempt to determine the force sufficient to keep the moon in its orbit. In another brilliant insight, he did not assume the Earth to be the lone actor in this celestial drama. Any object, he had come to believe, has the power to attract any other object. An apple attracts the Earth just as the Earth attracts an apple, although the planet's size is so great that the principle challenges common sense. And what is true of the apple is equally true of the moon.

Newton believed that the attraction between the Earth and the moon must decrease inversely as the square of the distance separating them. Hence, if this attraction amounts to a certain force when the bodies are at a given distance from each other, at twice the distance the force will not be half as great but one-fourth as great; at three times the distance, not one-third as great but one-ninth as great, and so on, in ever smaller proportions as the distance increases. The same formula must apply to the sun and planets, suggesting that a universal principle is at work. As Newton himself later wrote of this period, "I deduced the forces which keep the planets in their orbs must be reciprocally as the squares of their distances from the centers about which they revolve."[6]

But there were problems. Try as he might, the young genius could not get his calculations to work out as he had expected. According to the mathematician William Whiston, in whom Newton later confided, "Upon this disappointment . . . Sir Isaac . . . threw aside the paper of his calculation and went on to other studies." "To be looked upon as one of my guesses," Newton joked years later after finally achieving the result he was after.[7] Ah, but what a "guess" it was!

Though relatively secure from the plague, which had begun to subside, Newton remained at Woolsthorpe throughout most of 1666, where, in September, word arrived that another tragedy had befallen the beleaguered citizens of London. The Great Fire, as it came to be known, began on September 2, and raged out of control for four days and four nights. The flames laid waste an area a mile and a half long and half a mile wide, consuming 436 acres, more than 13,000 houses, and 87 churches, including beautiful old St. Paul's Cathedral. The remains of the city were leveled as though on a great battleground; a deep and unbroken sea of ashes, soot, and dirt were all that remained of the London beloved by Shakespeare and Elizabeth I. Many believed that the fire, like the Great Plague, was an act of divine retribution brought on by a God made angry by the sinful ways of his children, a view that

Newton, with his strong puritanical streak, may well have shared. There was, however, at least one important blessing to be counted in the fire's aftermath. While tens of thousands had died from the plague, the death toll from the fire was miniscule by comparison. The official Bills of Mortality listed only six people as having perished in the flames.

One casualty of the plague was Sturbridge fair, which Newton had visited sometime before the epidemic forced its closing. It was there, he wrote, that "I procured me a triangular glass prism to try therewith the celebrated phenomena of colors."[8]

The prism had been known since the time of the Roman philosopher and statesman Seneca in the first century A.D. But even before this, Aristotle had argued that colors are a mixture of light and dark, or of black and white, a view that was still accepted by most natural philosophers of Newton's day. It was thought that red was the strongest, least modified, and closest to pure white of all the colors, while blue was the weakest, most modified, and closest to black. But no matter what the color from the spectrum—whether red, orange, yellow, green, blue, indigo, or violet—all were believed to be modifications of pure white light.

At Woolsthorpe, Newton conceived of a novel experiment. Darkening his upstairs bedroom, which faced south, he drilled a one-eighth-inch hole in a shutter and intercepted the beam of incoming light with a prism. The resulting spectrum was projected on the wall across the room, which served as a screen. To his great surprise, the spectrum took the form of an oblong strip rather than the perfect circle Descartes said should appear. This could mean only one thing: each of the seven colors was refracted at different angles on passing through the prism—red the least and violet the most, and the other colors in between.

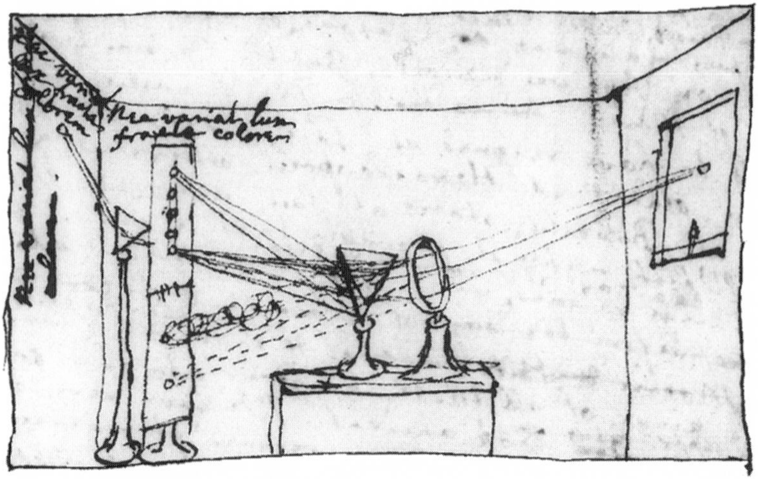

Newton's drawing of his so-called crucial experiment shows light from the sun being refracted through one prism and then being refracted through another prism. The colors never changed. *Bodleian Library, Oxford*

Ever cautious, Newton conceived of a second, equally powerful, experiment that he later termed the *experimentum crucis,* the crucial test. After placing a prism near the drilled hole, he intercepted the spectrum with a second prism located about "5 or 6 yards from ye former."[9] As with the first prism, blue rays were bent at a greater angle than the red ones. "The fashion of the colors was in all cases the same," he wrote.[10] He then realized that if the angles of the bent colored rays could be accurately calculated, a new law of refraction could be formulated. He quickly set about determining these angles, or sines, a further vindication of his vision that nature operates according to precise mathematical principles.

The *experimentum crucis* led to another, equally powerful, insight. Newton observed each ray as it passed through the second prism. Just as he suspected, blue remained blue, orange remained orange, and so forth. He rotated the prism and still each color remained the same. If colors were nothing but modifications of white light, as had long been believed, the second prism should have produced other colors by turning red to orange or blue to indigo, but it did not. White light, he had proven beyond any reasonable doubt, is made up of all the colors of the spectrum, none of which are altered on passing through a prism. In the language of a poet, Isaac Newton had "untwisted the shining robe of day" and deftly put it back together again.

Looking back over Newton's achievements more than three centuries later, historians have termed the Woolsthorpe period the *annus mirabilis*—the wondrous year—though the discoveries actually took place over the course of two years rather than one. So far as we know, Newton had not yet communicated anything of what he had learned to another human being. But perhaps as a way of rewarding himself and pleasing his mother, he traveled the familiar road to Grantham in 1666. There before the king's herald, he signed himself "Isaac Newton of Woolsthorpe, Gentleman, age 23."[11]

Four

The Revolutionary Professor

With a bachelor's degree in hand and his status as a gentleman legally recorded, Newton returned to Cambridge and Trinity College to complete his master of arts. He now faced the fellowship elections of September and October 1667. If successful, he could extend his stay at Trinity indefinitely; if not, he would likely be forced to choose between the life of an obscure gentleman farmer or the pastorship of a village church in rural Lincolnshire.

For three days during late September, Newton and his fellow candidates assembled in the chapel to undergo oral questioning. On the fourth day each of the nine was given a theme by the master of Trinity to be completed in writing within a period of six hours. As when Newton sat for bachelor's degree exams, the subjects covered were mostly drawn from the writings of ancient Greek and Roman scholars. On October 1, "by ye tolling of ye little bell," the successful candidates were called back to the chapel. The bell tolled for Newton, who, together with his trusted companion John

Wickins, swore to embrace the "true religion of Christ." Now officially a Fellow of the college, Newton had been granted permanent citizenship in the academic community. Never again would he have to face the unsettling prospect of being wrenched from the labors for which he was so admirably suited.

Together with Wickins, Newton celebrated—one of the few times in his life that he did so. The two hired a painter and a carpenter to refurbish their living quarters and then ordered new carpets and furniture, including a couch and "ticking for a featherbed." Newton also took an interest in his appearance and spent heavily on suits and shoes during the next two years. Added to this was the cost of a bachelor's gown, followed by an even larger expenditure for a tailor-made master's gown a year later. As was the custom, Newton shelled out seventeen shillings to celebrate his B.A. and spent fifteen more when he became master of arts on July 7, 1668. In between he recorded being "at the tavern several other times," where he even lost fifteen shillings playing cards. Then, after entertaining his "cousin Ayscough" and an unidentified acquaintance, these expenditures ended as abruptly as they began.[1] By all indications, Newton had returned to his solitary ways.

Not until August 1668, a month after he received his master's degree, did Newton first visit London. Although reconstruction was well under way, nearly four-fifths of the city still lay in ruin, a blackened monument to the massive devastation wrought by the Great Fire. Where he stayed and how he occupied his seven weeks in the capital remain minor mysteries, but he likely purchased books and supplies for his small laboratory and perhaps met certain mathematicians and natural philosophers whose works had inspired his scientific research.

Thus far the only scholar who knew anything of Newton's brilliance was Isaac Barrow, Lucasian Professor of Mathematics at Cambridge. This unusual academic had traveled throughout the Mediterranean and the Middle East, besting a "loose-tongued Turk" in hand-to-hand combat and bravely repelling an attack on his ship by Maltese pirates. A preacher by training if not necessarily by temperament, a heavy smoker, and slovenly in dress, Barrow nurtured a keen interest in optics, analytical geometry, and the mechanical model of the universe.

Newton recalled having been in the audience when Barrow gave the first of his Lucasian lectures in 1664, and allowed that Barrow may have stimulated his interest in advanced mathematics, but nothing more. Yet, it's possible that Newton visited Barrow in his quarters during the weekly hours set aside for discussing matters raised in his lectures, and that Barrow lent the student books from his personal library. Yet Stukeley tells us that during the scholarship examinations of April 1664, Dr. Barrow found that Newton had neglected Euclid and "conceived then but an indifferent opinion of him."[2]

Whatever the nature of their early relationship, by 1669 Barrow was well aware of Newton's unique mathematical gifts. Upon receiving of a copy of Nicholas Mercator's new treatise, *Logarithmotechnia* (The Art of Logarithms), from the London mathematical intermediary John Collins, Barrow drafted an enthusiastic reply, which struck Collins like a bolt from the blue: "A friend of mine here, that hath a very excellent genius to these things, brought me the other day some paper, wherein he has set down methods of calculating the dimensions of magnitudes like that of Mr. Mercator concerning the hyperbola, but very general."[3] The unidentified friend, of course, was Isaac Newton, and the paper referred to was his little treatise *De Analysi per Aequationes Infinitas* (On Analysis by Infinite Series), which Barrow promised to forward in his next letter.

It was with much difficulty that Barrow persuaded Newton to present his work to the scholarly world. This seems all the more strange because Newton's very purpose in drafting *De Analysi* was to establish his rightful priority to the method published by Mercator. An anxious Collins had to wait two weeks before he received the promised material, and still Newton's name was not revealed. Only after Collins replied in the most enthusiastic terms did Barrow respond: "I am glad my friend's paper giveth you so much satisfaction. His name is Mr. Newton, a fellow of our college, and very young . . . but of an extraordinary genius and proficiency in these things."[4] Barrow, with Newton's permission, allowed Collins to show the paper to Lord Brouncker, president of the Royal Society, and additional copies were made for circulation among several of Collins's associates. Under no circumstances, however, would the naturally secretive Newton allow a frustrated Collins to publish his work. Not until 1711, when Newton was sixty-eight, did *De Analysi*, one of the earliest examples of his revolutionary fluxions, reach print.

The belief that Barrow had been keeping an eye on the budding Newton for some time is further borne out by the Lucasian professor's request that Newton help him edit his forthcoming book on optics. Whether Barrow was aware of Newton's recent discoveries in this field remains another mystery. Since his own work was still in the experimental stage, Newton may have chosen not to dampen Barrow's spirits by discussing experiments that, if confirmed, would render the professor's work obsolete. Moreover, Newton was in quest of a major plum and wanted to do nothing to jeopardize his chances. Barrow had always considered himself to be more of a theologian than a natural philosopher, so when the opportunity arose to become chaplain to King Charles II, he jumped at it. Shortly before Barrow's departure for London, and doubtless as a result of his considerable influence, Isaac Newton was appointed Cambridge's second Lucasian Professor of Mathematics on October 29, 1669.

Newton was only twenty-seven years old, but his shoulder-length hair was already turning a beautiful silver. Wearing the scarlet robes of his new position, he strode across Trinity's Great Court to deliver the first of his lectures in early 1670, no doubt turning some heads along the way. Free to choose among a broad range of subjects, he selected optics, his first and greatest love among the experimental sciences. "I judge," he began, that "it will not be unacceptable if I bring the principles of this science to a more strict examination."[5]

The work with prisms that Newton had undertaken during the mid-1660s had been refined with a passion put to use. As reported by the Scottish physician Dr. George Cheyne, when Newton was carrying on his investigations into light and colors, "to quicken his faculties and fix his attention, [he] confined himself to a small quantity of bread, during all the time, with a little sack and water, of which he took as he found a craving or failure of spirits."[6]

Three years' additional labor now enabled Newton "to bring forth [his] opinion more distinctly."[7] This he did in a series of eight lectures delivered before an audience that dwindled rapidly after his first address. Indeed, none of his colleagues or students at Trinity left a record of having been in attendance, a symptom of the general lack of interest in the new science. Even Barrow, a popular figure on campus, had taught only a handful of followers during his five years in the Lucasian chair, sometimes lecturing to an empty room. If Newton's revolutionary discovery that sunlight is a mixture of all the rainbow colors was to reach a wider audience, he would have to communicate it to a world beyond Trinity's protective walls.

Among the books he had acquired as a young man was a copy of *Optica Promota* (The Promotion of Optics) by James Gregory, the distinguished Scottish mathematician and astronomer. As was his habit, Newton dog-eared the passages most important to him, including Gregory's design for the reflecting telescope. As yet, no one had successfully built a working reflector. In 1663 Gregory himself had gone so far as to order a special lens from Richard Reive, London's leading instrument maker, but Reive had failed in his attempts to produce the delicate object.

Newton took up the challenge a few years later, and his reason for doing so is not difficult to understand. Unlike the refracting telescope, which forms an image of an object with lenses, the reflector contains a parabolic mirror from which all light is reflected at the same angle. The advantage of this design is that the observer is not hindered by chromatic aberration, the hazy, rainbow-like phenomena produced when rays of different wavelengths are not brought to the same focus on passing through a lens. Though no one but Newton realized it at the time, the reflecting telescope, in producing an image free from distortion, offered further support for his theory of light and colors.

With the same skilled hands that had produced the models and clocks of his youth, Newton set to work. Using a thin piece of metal rather than glass, which is almost impossible to grind evenly with hand tools, he created a hollow in the shape of a saucer. He then prepared a special alloy composed of copper, tin, and arsenic, which is white in color and takes a high polish. When this tedious task was completed, he coated the metal mirror, or speculum, and placed it and other components in a small tube. In a letter to an unnamed friend, dated February 23, 1669, he described its performance. Newton calculated that the instrument was capable of magnifying "about 40 times in diameter which is more than any 6 foot tube can do, I believe with distinctness. I have seen with it Jupiter distinctly round and his satellites." He had no doubt that a carefully crafted six-foot reflector would perform as well as any "60 or 100 foot tube made after the common way." Realizing how puzzling this claim might sound, he added, "It may seem a paradoxical assertion, yet it is the necessary consequence of some experiments which I have made concerning the nature of light."[8]

Word of Newton's triumph spread to London, where he was erroneously hailed as the inventor of the reflector. This circumstance was reminiscent of 1609, when Galileo constructed the first refracting telescope in Italy from a description sent to him by a scientific correspondent from abroad. Most anxious to see Newton's telescope were members of the Royal Society, a scientific order founded in 1660 and chartered by none other than King Charles II. The society took as its motto part of the Latin quotation *Nullius in verba*, which freely translates, "Not by word of mouth." Its membership, which included the finest scientific minds of the time, both English and foreign, were pledged to the experimental method as opposed to the ancient practice of observation alone. The society's findings were regularly published in the renowned *Philosophical Transactions*, which became the model for other scientific journals—even to the present day.

This time Newton found the urging of his peers impossible to resist. He constructed a new, somewhat modified version of his first reflector and turned it over to Barrow, who proudly carried the little instrument to London at the end of 1671. It created a sensation and was soon brought to Whitehall, where an anxious King Charles II was treated to a demonstration of its powers. Meanwhile, Henry Oldenburg, the secretary of the Royal Society, wrote the great Dutch scientist Christiaan Huygens (1629–1695) to inform him of Newton's success. On receiving word of the reflector, Huygens wrote back, describing the new instrument as nothing less than "the marvelous telescope of Mr. Newton."[9]

Oldenburg also wrote Newton a letter of congratulations in which he informed the secretive professor that he had been nominated for membership in the Royal Society. Newton gratefully acknowledged the honor, then dropped a bombshell: "I am [proposing] and considering to be examined an account of a philosophical discovery which induced me to the making of the said telescope, and which being in my judgment the oddest if not the most considerable detection which hath hitherto been made in the operations of nature."[10]

Little did anyone in London realize that Newton was referring to his discovery that white light is composed of the primary colors. Yet once he had so boldly promised to reveal one of nature's greatest secrets, he suddenly began having second thoughts. "I hope," he wrote in his next letter to a puzzled Oldenburg, "I shall get some spare hours to send you . . . that account [of light] which I promised."[11]

After wavering, Newton wrote up the major steps in his optical discoveries that led him to the *experimentum crucis*, and even included an explanation of colors of the rainbow. Water droplets that refract the most light rays are on the exterior of the rainbow and appear to be red, while droplets comprising the inner layers deflect less light and appear as darker colors. He concluded by requesting that if any of the Fellows were interested enough to repeat his experiments, "I should be very glad to be informed with what success."[12]

Newton, it seems, need not have worried. Oldenburg placed his paper on the agenda of the next meeting, where it was read aloud. The session had no sooner adjourned than the secretary wrote him to report the result: "[The] reading of your discourse concerning light and colors was almost their only entertainment for that time. I can assure you, Sir, that it there met with a singular attention and uncommon applause."[13] Oldenburg urged Newton to give him permission to publish the paper in the next issue of the *Philosophical Transactions*, which would bring the author lasting fame.

Newton's spirits soared when he received word of the reception of his work. As to the printing of the paper, he replied to Oldenburg, "I leave it to their pleasure," a far cry from his earlier refusal to allow John Collins to publish the equally revolutionary *De Analysi*.[14]

Five

"Kindling Coal"

As curator of experiments for the Royal Society, Robert Hooke was given the considerable task of confirming Newton's findings with prisms. No portrait of Hooke survives, but the diary writer Samuel Pepys observed that he "is the most and promises the least of any man in the world I ever saw."[1] Of medium stature, Hooke suffered from a bent spine; his large head, dominated by bulging gray eyes, seemed too big for his body. A pale complexion and fixed stare created the impression of one detached and unaware of his surroundings. Taking pity on his misshapen friend in print, the biographer John Aubrey felt compelled to add of Hooke, "He haz a delicate head of haire, browne, and of an excellent moist curle."[2]

Despite his looks, Hooke possessed intellectual gifts that bordered on genius and in temperament was cast in a mold reminiscent of Leonardo. His thought process, like that of the great

Florentine, was an erratic mixture of sudden uprisings and agonizing downfalls, of crescendo and diminuendo. He raced from problem to problem, often attacking new ones before resolving previous questions. Indeed, it is on men like Hooke that fate plays one of its cruelest jokes, blessing them with great intuition while denying them the mathematical gifts to transform their visions into concrete principles. Always painfully conscious of more than he could prove, Hooke watched in agony while others reaped the harvest for which he had helped prepare the ground.

Like Achilles and Hector, Newton and Hooke seemed destined by the gods to do battle. On the day Newton's reflecting telescope was first examined by the admiring Fellows of the Royal Society, a jealous Hooke argued that the invention was of little moment. Hooke claimed he had fashioned a refracting telescope eight years earlier that was only an inch long and could be worn on the chain of his pocket watch. According to its maker, it performed better than any telescope—even one fifty feet long. Only the onset of the plague had kept him from constructing a larger version of the instrument. He had also wanted to keep his secret from the glass grinders. At least some of those present must have been skeptical, for Hooke liked to brag and often claimed that the fear of theft was the reason for not making his inventions and discoveries known.

Ever in a hurry, Hooke later admitted that he spent very little time reading Newton's paper and duplicating his experiments, a step essential to confirming the results. Yet he agreed with Newton's conclusions, with one major exception. After stating that he would "not mingle conjectures with certainties," Newton had asserted that "it can no longer be disputed whether light be a body."[3] In other words, experiments had led him to embrace what was then known as the "corpuscular theory," which held that a light beam is composed of minute particles, or corpuscles, whose "blows" on the surface of objects produce the colors of the spectrum. Conversely, Hooke, who leaned toward the theory that light is composed of waves rather than of tiny particles, took Newton to task for making an unproven claim: "Nay, even those very experiments which he alleged do seem to me to prove that light is nothing but a pulse or motion propagated through a . . . uniform and transparent medium."[4]

Angry at having left himself open to criticism, Newton delayed his reply, vowing never to make the mistake of mixing fact with conjecture again. Meanwhile, comments began coming in from other natural philosophers who had read his paper in the *Philosophical Transactions*. Foremost among them was the great Dutch scientist Christiaan Huygens, who had taken up residence in Paris after being promised a handsome pension by Jean-Baptiste Colbert, minister of finance to Louis XIV, the Sun King. Having earlier praised Newton's telescope, Huygens pronounced the new theory of colors "highly ingenious." Coming from Europe's greatest natural philosopher, this was heady praise indeed.

But there were rumblings from other quarters that Newton did not appreciate. Sir Robert Moray, who served as the first president of the Royal Society, proposed four simplistic and essentially worthless experiments as a means of verifying the *experimentum crucis*, which Moray lacked the skill to duplicate himself. Then from France came a long letter written by Ignance Gaston Pardies, a professor of speech and a member of the Society of Jesus, a learned order of Roman Catholic priests. Like Moray, Pardies had failed in several attempts to duplicate the *experimentum crucis* and concluded that Newton's results lacked scientific merit. A deeply irritated Newton drafted a sharp reply in which he concluded that "[my results] are difficult to prove, and which if I did not know to be true, I should prefer to reject as vain and empty speculation, than acknowledge them as my hypothesis."[5]

A shaken Pardies next wrote to Oldenburg. He included a comment for Newton, but begged the Royal Society secretary to examine its contents for anything offensive before forwarding it to Cambridge. Pardies had withdrawn his first argument but raised another in its place, which again challenged Newton's skills as an experimentalist. Newton sent an even more chilling blast across the English Channel, scolding the Jesuit for his sloppy work and for "philosophizing" instead of "establishing properties by experiments." At Newton's instruction, Pardies took up his prisms once again and this time reproduced the very results contained in Newton's original paper. "The last scruple which I had about the *experimentum crucis*," Pardies wrote Oldenburg, "is fully removed. I now clearly perceive by [Newton's] figure what I did not before understand. . . . I have nothing further to desire."[6]

Newton had no sooner settled his accounts with Pardies when Oldenburg began pressing him for a publishable reply to Hooke's criticism of the corpuscular theory. Alerted to Newton's temper, Oldenburg cautioned Newton to "aim at nothing but the discovery of truth" instead of attacking persons by name.[7]

It was a warning Newton chose not to heed in his response of June 11, 1672, in which Hooke is named at least thirty times. Newton began by expressing disappointment in a person from whom he had expected an impartial examination of his hypothesis. Once the knife was inserted, he began twisting the blade: "Mr. Hooke knows well that it is not for one man to prescribe rules to the studies of another, especially not understanding the grounds on which he proceeds."[8] He then went on to show that it did not matter which mechanical hypothesis one might embrace; whether corpuscles, waves, or some other, the principle of colors remained unchanged. Hooke would do well to repeat the *experimentum crucis* rather than rely on "hypothetical explications."

It was a brilliant, if not necessarily fair, exercise in turning the tables. Instead of asking Newton to redraft his response, Oldenburg, who disliked the know-it-all Hooke, immediately presented the reply to the council of the Royal Society. Hooke was humiliated and suffered additional embarrassment when Oldenburg published the text in the November issue of the *Philosophical Transactions*. Hooke's criticism of Newton's work never reached print.

Still, Newton must have felt a lot like the Dutch boy who attempted to plug holes in the dike with his fingers; each time one was sealed another opened up. In the autumn of 1672 Huygens, his only important ally in the international scientific community, expressed second thoughts about the conclusions on light in a letter to Oldenburg. After the letter was passed on to Newton, the secretary received a stunning reply: "Sir, I desire that you will procure that I may be put out from being any longer a Fellow of the Royal Society. For though I honor that body, yet since I see I shall neither profit them, nor can partake of the advantage of their assemblies, I desire to withdraw."[9]

Oldenburg and John Collins, who was still corresponding with Newton on mathematics, quickly put their heads together in hopes of averting disaster. Taking Newton's resignation for what it was—a plea for emotional support—Oldenburg not only promised to waive Newton's quarterly dues but wrote, "You may be satisfied that the [Royal Society] in general esteems and loves you, which I can assure you of."[10] Newton was not completely satisfied, but when he wrote back two weeks later he dropped the matter of his resignation. However, he wrote, "I must signify to you that I intend to be no further solicitous about matters of philosophy."[11] Needless to say, Oldenburg forwarded no more letters to him containing objections to the controversial paper.

A curtain of silence descended. For the next two and a half years Newton scarcely corresponded with anyone. Then, in September 1674, Francis Hall, or Linus, as he called himself in Latin, sent a letter to Oldenburg criticizing Newton's experiments. Oldenburg was hesitant to pass it on, but if he did not do so his position as an impartial scientific agent would be compromised. The letter was forwarded, and it did not take long to learn that the Cambridge professor remained as touchy as ever. Newton wrote Oldenburg that Linus, an eighty-year-old professor of Hebrew and mathematics, had gotten it all wrong, just as his fellow Jesuit Pardies had. Newton reaffirmed his earlier position that he was no longer interested in debating the finer points of natural philosophy, but if Linus was so foolish as to publish the results of his poorly conducted experiments he would only be "slurring himself in print with his wide conjecture."[12]

Meanwhile, during a rare visit to London, Newton had been surprised at the high praise lavished on his work by members of the Royal Society. Even the deformed Hooke seemed less of a threat when viewed in person. Indeed, Newton mistakenly thought that the curator of experiments had abandoned his previous notions concerning colors and was coming around to his point of view.

Newton, who secretly craved recognition, decided to break his vow of silence. Near the end of 1674, an elated Oldenburg received Newton's most daring and complex paper to date—"An Hypothesis Explaining the Properties of Light." The Fellows were so impressed that the meetings of December 9 and 16 were almost entirely devoted to its contents. Then, following Christmas recess, its many principles were deliberated during the meetings of December 30 and January 13, 1675.

The truly great scientific mind distinguishes itself in two ways: first, in its capacity for original thought; second, in setting forth a principle by which knowledge, both old and new, is forged into a grand design. Newton's work on gravity during the plague years had involved more than mathematical calculations. He was no less interested in determining the physical *cause* of this perplexing phenomenon, a subject to which he had devoted countless hours of thought during the intervening years. Having rejected hypothetical speculation on the part of others, he now found himself in the awkward but exhilarating position of defending a sweeping hypothesis of his own.

In his paper Newton postulates the existence of ether, a medium or agent by which many forces acting on matter throughout the universe are generated. Much less dense and more elastic than air, ether cannot be seen or felt. And while it is to be found everywhere, it is rarer within the dense bodies of the sun, stars, and planets than in the vast spaces separating them. In his mind's eye, Newton saw the Earth and the celestial bodies as giant sponges, steadily soaking up the stream of invisible matter that constantly presses down upon their surfaces. This stream, he wrote, "may bear down with it the bodies it pervades with force proportional to the surfaces of all their parts it acts upon," the first known musings on universal gravitation. Once the ether has penetrated into the "bowels" of a planet or star it is somehow transformed and returns to space where the process repeats itself in an endless cycle: "For nature is a perpetual circulatory worker, generating fluids out of solids, and solids out of fluids. . . . And as the Earth, so perhaps may the sun imbibe this spirit to conserve his shining and keep the planets from receding further from him, and . . . this spirit affords and carries with it thither the solar fuel and material principle of light."[13] Newton sought to explain gravity not only in terms of the ethereal medium but in terms of other equally puzzling phenomena. Such phenomena included cohesion, or the mutual attraction by which the various elements of a body are held together, and animal sensation, or what we now call feelings generated in the nervous system. Finally, he filled the paper with detailed accounts of many experiments designed to reinforce his soaring hypothesis.

Newton bolstered his theory in a second paper sent to London along with the first. Titled "Discourse of Observations," he speculated that when corpuscles of light pass through the ether, the medium's varying density alters the speed and direction in which the corpuscles move. This, in turn, produces the familiar effect of reflection as well as the scattering phenomenon called diffusion. Yet colors themselves do not arise from changes in the corpuscles. Instead, different colors are produced when the particles are separated from one another, creating the familiar spectrum.

Newton's second paper owed much to Hooke's own treatise on light, *Micrographia* (The Drawing of Small Objects), which contains an account of the colors of thin films such as soap bubbles. But where Hooke had relied exclusively on observation, Newton employed meticulous measurements and mathematical analysis, enabling him to take over ground first broken by the frustrated curator of experiments. In an attempt to save the day, Hooke called for a secret meeting with a few close friends at Joe's Coffee House in London, where, he later wrote in his diary, "we began a new club." The sole topic of conversation was Newton's latest work. After the group met again three weeks later, Hooke noted, "I showed that Mr. Newton had taken my hypothesis of the pulse or wave."[14] What was described as Hooke's gray eye had suddenly sprung to life.

The problem was not one of theft but Newton's failure to acknowledge Hooke's work. He would likely have won Hooke's support if only he had praised the credit-starved experimentalist as a source of fruitful ideas. Instead, Newton liked to think of himself as a wholly independent agent, a self-created myth his biographer Frank Manuel termed "the magic circle of his infallibility."

As suspicious as Hooke, Newton soon got wind of the secret meetings from Oldenburg and fired off a letter to London. "As for Mr. Hooke's insinuation that the sum of the hypothesis I sent you had been delivered by him in his *Micrographia*, I need not be much concerned. I may avoid the savor of having done anything unjustifiable or unhandsome towards [him]."[15] He again recounted the penetrating nature of his insights, which Hooke had barely glimpsed. To Hooke, Newton had become a fearsome rival; to Newton, Hooke was an intolerable nuisance.

When Oldenburg read some of Newton's passages criticizing Hooke at the Royal Society meeting of January 20, 1675, matters came to a head. The secretary had embarrassed Hooke once too often. Taking matters into his own hands, Hooke wrote directly to Newton in Cambridge about what he termed "Oldenburg kindling coal." "I do no way approve of contention or feuding or proving in print," Hooke began, "and shall be very unwillingly drawn to such kind of war."[16] He then praised Newton's "Hypothesis

Explaining the Properties of Light," judging that "you have gone much farther in that affair than I did."[17] In the future they would be better off corresponding directly with each other so as to produce light rather than heat.

Newton's reply was no less polite. He addressed Hooke as "his honored friend" whose "generous freedom" in writing "becomes a true philosophical spirit."[18] There was nothing he wished to avoid more than contention in print and gladly accepted Hooke's offer to correspond privately. In light of future developments, it is doubtful whether either was sincere. For the moment, however, the one-time country boys were acting like fawning courtiers—overpraising each other, bowing, scraping, tipping their plumed hats in grand gestures.

Four years had passed since Newton first informed the Royal Society of his experiments with prisms. Yet, difficult as it is to believe, neither Hooke nor anyone else had attempted to duplicate the results before the distinguished body. Finally, on Thursday, April 27, 1676, "a very clear sun-shine day," Hooke followed Newton's instructions to the letter, and the *experimentum crucis* was verified independent of the question of whether light is particle or wave, or whether ether is the mechanism by which gravity works. Hooke, who wrote down everything else he witnessed that day in his diary, including Thomas Tompion's repair of Lady Wilkins's watch and the direction of a drain through cellars, could not bring himself to record yet another of Newton's triumphs. But when word of the results reached Cambridge, Newton could not hide his pleasure. Surely now, he wrote Oldenburg, "Mr. Linus's friends will acquiesce."[19]

Six

The Alchemist

It was at the Globe Theatre, the London playhouse constructed in 1598, that most of William Shakespeare's dramas were first performed by a company of actors called Chamberlain's Men. Among the most memorable scenes played out on the Globe's stage was one from the tragedy *Macbeth*. It is set in a cave, where three witches circle a boiling cauldron, chanting:

Double, double toil and trouble;
Fire burn, and cauldron bubble.
Fillet of a fenny snake,
In the cauldron boil and bake;
Eye of newt and toe of frog,
Wool of bat and tongue of dog.

To an audience of Shakespeare's time, there was nothing unusual about witches or boiling cauldrons or the casting of spells over a vile brew of exotic ingredients. But when it later became known that Isaac Newton, who was born two years before the Globe burned down in 1644, also kept a cauldron stocked with odd ingredients, it was considered to be something of a scandal. After all, how could the greatest of natural philosophers have fallen victim to the practice of magic—the very superstition modern scientists were seeking to stamp out?

It is almost certain that Newton's introduction to the fire and the crucible came during his Grantham years when he lived with the apothecary William Clarke. Clarke prepared remedies for his customers in a small laboratory located either in or near his shop. That Isaac was a close observer is proven by the many recipes recorded in his early student notebooks, in addition to the volumes he began to collect for his personal library.

By the mid-1660s the philosophical notebook of Newton's Cambridge years was crammed with entries drawn from the works of a new group of experimenters in metals who called themselves "chymists." During an autumn visit to London in 1669, Newton purchased two portable furnaces and enough supplies to conduct scores of his own experiments. Though he was only in his late twenties, his long hair was now completely gray, prompting Wickins to remark that it was the "effect of his deep attention of mind." An unusually merry Newton replied that he had experimented so frequently with quicksilver, or mercury, "as from thence he took the Colour."[1]

Newton's laboratory adjoined his living quarters and could be seen from bustling Trinity Street a short distance away. By the 1680s he had become so deeply involved in experiments that he found it necessary to employ an assistant, a shirttail relative named Humphrey Newton who hailed from Grantham. Humphrey left a vivid portrait of his employer, whom he described as a man transported by his work:

> So intent, so serious upon his studies that he eat very sparingly, nay ofttimes he has forgot to eat at all. . . . He very rarely went to Bed till 2 or 3 of the Clock, sometimes not till 5 or 6, especially at spring & fall of the Leaf, at which Times he used to imploy about 6 weeks in his Elaboratory, the fire scarcely going out either Night or Day, he sitting up one Night, as I did another, till he had finished his Chymicall Experiments, in ye Performance of which he was the most accurate, strict, exact.[2]

But what, precisely, was Newton after? Of that, Humphrey had no inkling:

> What his Aim might be, I was not able to penetrate into, but his Pains, his Diligence at those set Times made me think he aimed at something beyond the Reach of human Art & Industry. I cannot say I ever saw him drink, either Wine, Ale or Beer, excepting Meals, & then but very sparingly.[3]

Newton, it seems, had succumbed to the ancient call described by one Ruland the Lexicographer: "The Alchemical meditatio is an inner dialogue with someone who is invisible, as also with god, or with oneself, or with one's good angel."

The truth is that Newton was neither a grand magician nor a modern chemist, but something in between. For many centuries learned men the world over had been seeking to solve a riddle as ancient as the pyramids. Calling themselves "alchemists," a word coined by the Arabs centuries earlier to describe those who worked with chemicals, they risked everything, including money, health, and the welfare of their families, in quest of the so-called philosopher's stone. The "stone," which was also called the "grand elixir," the "grand magistery," and many other names, was not a solid substance but the rarest of liquids possessed of legendary powers. When mixed with common metals, such as lead, it could transform them into the most precious substances—namely, silver and gold. If drunk on a regular basis, the stone promised immortality, such as that sought in the mythical Fountain of Youth by the sixteenth-century Spanish explorer Juan Ponce de León. Medieval alchemists couched their laboratory notes in cryptic symbols to protect their secrets from the prying eyes of rivals. Many gained the support of wealthy patrons who dreamed of great riches once the secret of transformation became known.

Strange as it may seem, Isaac Newton was perhaps the greatest alchemist of all. At his death he left hundreds upon hundreds of handwritten pages on alchemy, in addition to a remarkable document titled *Index Chemicus* (The Chemical Index). The "Index" contains 879 headings, with at least 5,000 references to information found in scores of other alchemical texts. In addition to his countless experiments, Newton read everything alchemical that he could buy, beg, or borrow, fearing that the smallest overlooked clue might conceal the key to the universal matter that would explain the very structure of the world. Dr. Thomas Pellet, a respected member of the Royal Society chosen to decide which of Newton's papers to publish after his death, was so horrified by the venture into alchemy that he wrote in bold letters across the wrappers: "Not fit to be printed."[4]

A closer examination of the alchemical papers by modern scholars has led to some startling, and rather more reassuring, conclusions. It is now clear that Newton was not primarily interested in getting rich or in prolonging his life, though he sincerely believed that such things were possible. Instead, he had to know everything there is to know about the behavior of matter, from the smallest particle to the grandest star. Before 1666 he had focused on what scientists now refer to as the macrocosm, or the cosmos itself. After 1666 his interests broadened to include the microcosm, the smallest worlds of invisible matter through which all things in nature are formed, grow, decay, and eventually return to their basic elements. By studying these worlds, Newton believed that he could discover what light truly is, how forces such as gravity and magnetism act across great distances, and how the theoretical ether of his experiments triggers changes in the bodies it inhabits.

In one sense Newton was preoccupied by the very ideas that fuel current scientific debates. Black holes, formed by collapsed stars at the centers of galaxies, are fascinating partly because their explanation promises to aid in unifying the large and the small. The two major theories of twentieth-century physics are relativity, which applies to light as it streaks across the great expanses of trackless space, and quantum mechanics, which seeks to comprehend invisible worlds of micromatter. The great problem is to forge a principle that combines both realms—the vast and the infinitesimal. Newton was the first to attempt it; Einstein was the second. Neither was successful and both men, despite their marvelous achievements, went to their graves disappointed at failing to do so. "Something beyond the reach of human Art & Industry," as Humphrey Newton had written.

Yet, in another sense Isaac Newton was less of a modern thinker than many would like to admit. Alchemy was not yet the science of chemistry, nor would it become so until the emergence of a brilliant generation of eighteenth-century experimenters. Their company included Joseph Priestley, Henry Cavendish, and, most important, the Frenchman Antoine Lavoisier, who distinguished chemical elements from the compounds long employed by alchemists. Moreover, Newton's alchemy was as deeply rooted in the past as in the present.

As an avid reader of old texts, Newton believed in a concept called the *prisca sapientia*, or wisdom of the ancients, a doctrine popular among certain scholars during the Renaissance. This doctrine held that the great truths of nature had been known to some of the most brilliant and morally upright thinkers of the distant past. Such thinkers included the Greek philosophers Plato and Democritus, the medieval scholars Albertus Magnus, Arnold of Villanova, and Roger Bacon, as well as some of the most important figures from the Bible—King Solomon, the prophet Isaiah, and Jesus of Nazareth. Not surprisingly, these men had hidden their knowledge of how nature works in symbolic language that, like a complex code, was extremely difficult to decipher. Newton took special note of a passage from the Book of Isaiah in which the ancient Hebrew promises enlightenment to the true believer: "And I will give thee the treasures of darkness, and hidden riches of secret places, that thou mayest know that I, the Lord, which call thee by thy name, *am* the God of Israel."[5] In one of his many manuscripts, Newton echoes the prophet's words by arguing that wisdom and insight are "not only to be found in the volume of nature but also in the sacred scriptures, as in Genesis, Job, Psalms, Isaiah & others. In the knowledge of this Philosophy God made Solomon the greatest philosopher in the world."[6] Little wonder that Newton drew up detailed plans of the great temple Solomon constructed in Jerusalem in hopes of finding clues to the secrets of nature in its design.

There is no doubt that Newton considered himself specially chosen to rediscover and expand the ancient wisdom, thus becoming one of the few from each generation so blessed. After all,

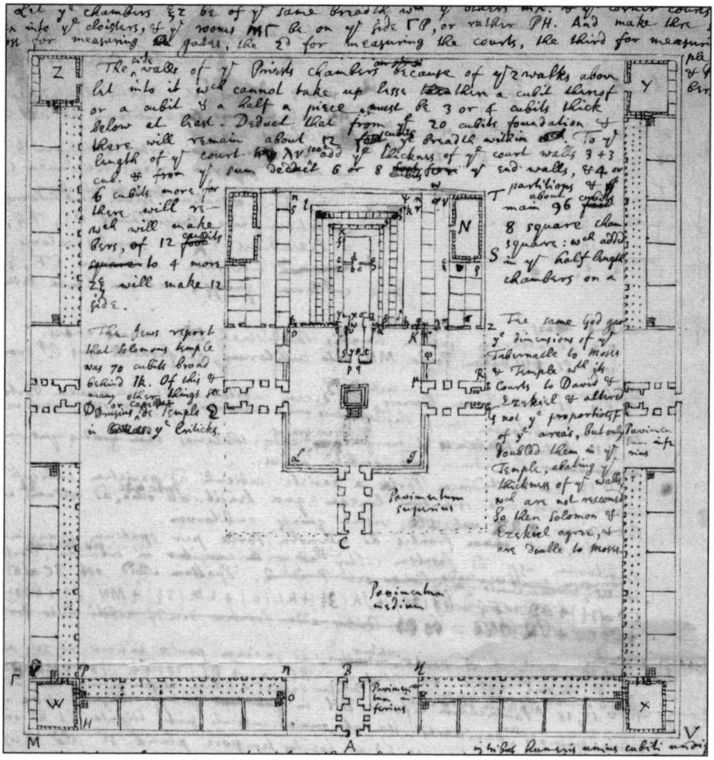

After a careful reading of the Bible, Newton was able to reconstruct this plan of the temple built by King Solomon in Jerusalem. *Burndy Library, Cambridge, Massachusetts*

he was born on Christmas Day and had survived his premature entry into the world when everyone, except perhaps his mother, thought he would die within hours of his birth. He rose to the highest place at Grantham's King's School and then, during the "miraculous year," unraveled the secret of light, gave birth to a revolutionary mathematics, and undertook calculations that would soon elevate his musings on gravity to the realm of scientific truth.

Who else among his generation had accomplished so much, and at such an early age? And why should his pursuit of alchemy not reveal even more secrets about the behavior of matter and its effect on everything from atoms to the stars?

Besides containing invaluable keys for unlocking what Newton termed nature's "operations," Scripture to him served as both a moral guide and a means of determining what would happen in the future. The earliest clues to his private thoughts on God are contained in the confession of sins he penned at Cambridge when he was nineteen: "Not turning nearer to Thee for my affections. Not living according to my belief. Not loving Thee for Thy self."[7] From these simple sentences he would go on to write an estimated 1,400,000 words on religion, more than the alchemy, more than the mathematics, more even than the physics and astronomy that made him immortal.

Newton came to know the Old and the New Testaments better than most clergymen. And, as with everything else he studied, his interpretation of Scripture differed radically from that of others. Unlike Wickins and his fellow scholars at Trinity College, Newton was not ordained into the ministry of the Church of England. With Isaac Barrow's assistance, he successfully petitioned King Charles II to set aside the legal requirement that he take the oath of ordination upon graduating.

By the 1670s Newton had become convinced that a major principle embraced by the church was wrong. The doctrine of the Trinity, to which his very college had been dedicated by Henry VIII, was in his eyes a form of heresy. Newton rejected this idea on the grounds that the worship of three figures in one violates the First Commandment, which states that there shall be only one God, the Creator of the universe. To accept ordination, or Holy

Orders, would be to sacrifice his conscience, something he would not do.

Yet, at the same time, Newton was being less than honest when he petitioned the king, for he kept his real reason for not taking Holy Orders secret. Why he chose to lie is easily understood: if his rejection of church doctrine was ever discovered, he would be dismissed from Trinity College and forced to return home to Woolsthorpe in disgrace, his chances of ever making a name for himself ruined. Instead, he kept his explosive opinions to himself, and only on occasion revealed them to an inner circle of young admirers once he became a famous man.

Employing sacred texts, Newton compiled a dictionary of religious terms, events, and images in a manner reminiscent of the *Index Chemicus*. He believed that biblical prophecy, especially as set forth in the Book of Daniel in the Old Testament and the Revelation of St. John in the New, provided an excellent chronology of events to come. Of course, the writings required interpretation and once again Newton thought he was just the person for the job: "Having searched after knowledge in the prophetique scriptures," he wrote while he was still in his thirties, "I have thought myself bound to communicate it for the benefit of others."[8] Such communication would not take place until after his death, however, when his *Observations upon the Prophecies of Daniel, and the Apocalypse of St. John* was published in 1733. In this work he predicted that Satan's spell would be broken and that the "fearful, and unbelieving, and the abominable, and murderers, and sorcerers, and idolators, and all liars shall have their part in the lake which burneth with fire and brimstone."[9] And just exactly when would this come to pass? Never shrinking from a challenge, Newton calculated that it would occur in the year 2060.

It is easy to dismiss Newton's religious beliefs as little more than the eccentric musings of an otherwise supremely rational mind. But it must be remembered that however modern his science may seem to us, Newton was still a man from the seventeenth century, a man of his time. Unlike many thinkers today, he saw no conflict between science and religion and wrote that the world could not operate without God being present. Indeed, without the Creator's periodic intervention the universe would eventually run down, only to collapse and explode as the planets, comets, and stars rush together as part of the final cataclysm long foretold in prophecy. Few things would have angered or dismayed Isaac Newton more than the claim by a later generation of thinkers that his formulation of mechanical laws established the framework of a universe in which God is no longer a vital, or even necessary, part.

Seven

"A Book Nobody Understands"

In August 1684 a handsome young astronomer named Edmond Halley boarded the London coach for Cambridge and sat back to ponder the events that had sent him on an important mission. Earlier in the year, he had entered into a lively conversation with Robert Hooke and Sir Christopher Wren, noted architect of the new St. Paul's Cathedral in London. Halley suggested that the force of attraction between the planets and the sun decreases in inverse proportion to the square of the distance between them. If this were true, then each planet's orbit should take the form of Kepler's ellipse, a shape like a football, though somewhat more rounded.

Halley recalled that Hooke immediately "affirmed that upon that principle all the Laws of the celestial motions were to be demonstrated."[1] Wren, who was also deeply interested in the new science, claimed that he, too, had reached the same conclusion. The problem, as all three admitted, was to find the mathematical means of proving it.

Anxious for a solution, Sir Christopher offered to give a valuable book to the friend who could deliver solid proof within the next two months. Hooke, to whom modesty was a stranger, claimed that he already possessed the required proof. However, he would keep it a secret for the time being so that his friends "might know how to value it, when he should make it public."[2]

The deadline came and went without a word from Hooke, and spring soon turned to summer. Finally, after seven months of silence, Halley decided to act. He cast an anxious eye in the direction of Cambridge and made a fateful decision. He would visit Trinity College to see if the secretive Isaac Newton could shed some light on the matter.

Newton was living an even more isolated existence than before. Some years earlier his mother, Hannah, had caught what was described as a "malignant fever," a catchall term for any number of fatal illnesses. Newton hurried to Woolsthorpe and took charge of Hannah's care, dressing her blisters and sitting up all night at her bedside. Unfortunately, she was beyond saving and died some days later. As her first child, Isaac inherited most of her property, making him an independently wealthy man.

A more recent blow was the departure from Cambridge of John Wickins, Newton's roommate of twenty years. Wickins became minister of the parish church at Stoke Edith, married, and fathered a son named Nicholas. Though they had been through much together, the two friends would never meet again and exchanged no more than a letter or two in the coming years.

Newton's prescription for loneliness was work, work, and more work. Humphrey Newton observed: "I never saw him take any Recreation or Pastime, either in Riding out to take the Air, Walking, Bowling, or any other Exercise whatever, Thinking all Hours lost that was not spent in his Studies, to which he kept so close that he seldom left his Chamber."[3] The reclusive professor had even more time to himself because Cambridge students were little interested in natural philosophy. Humphrey noted that his employer often lectured to the classroom walls. Finally, he stopped going to the lecture hall altogether, placing his texts on deposit in the college library as was required by statute.

Over the years, Newton became the very model of the absent-minded professor. He ate little and often had to be reminded by Humphrey that the dinner delivered to his room had gone untouched. He would express surprise, walk over to the table, and eat a bite or two standing up. "I cannot say," Humphrey noted, "I ever saw Him sit at [the] Table by himself."[4]

Newton rarely went to bed until two or three in the morning and often slept in his clothes. He rose at five or six, fully refreshed. His long, silver hair was seldom combed, his stockings hung loose, and his shoes were down at the heels. On those rare occasions when he went out, it was usually to take a meal in the dining hall, overlooked by the giant portrait of Henry VIII.

But, according to Humphrey, things did not always go as they should have. Instead of heading almost straight across the Great Court, Newton sometimes made a left turn, only to wind up on Trinity Street. On realizing his mistake, he would turn back, and then "sometimes without going into the Hall return to his Chamber again."[5]

In good weather Newton was occasionally seen taking a stroll in his garden. Picking up a stick, he drew figures on the graveled walks, which the other Fellows sidestepped for fear of ruining a work of genius—or folly. According to Humphrey, "When he has some Times taken a turn or two he has made a sudden stand, turn'd himself about, run up the Stairs like another Archimedes, and with a eureka fallen to write on his Desk standing, without giving himself the Leisure to draw a chair to sit down on."[6] So absorbed was Newton that he lost track of time and place. The days and dates on many of the papers recording his experiments do not match those of the calendar.

As he stepped down from his coach on reaching Cambridge, Halley hardly knew what to expect. He had exchanged no letters with Newton and had met him only once before, in London. Furthermore, Hooke's name was bound to come up. Despite their mutual pledge not to kindle any more coal, Newton and Hooke were still feuding over scientific matters both great and small.

To Halley's surprise, and great relief, Newton was flattered by his visit. They talked of many things before the astronomer revealed his reason for seeking Newton out. What kind of curve, Halley finally asked, "would be described by the planets supposing the force of attraction towards the sun to be reciprocal to the square of their distance from it?"[7]

Without hesitation, Newton responded that it would be an ellipse! Taken aback, Halley asked him how he knew it.

"I have calculated it," Newton replied.[8]

When Halley asked to see the calculations, Newton began rummaging through his many stacks of papers while his excited visitor held his breath. As luck would have it, he was unable to find the critical documents, forcing Halley to depart without the written proof he required. However, all was not lost. Before they parted company, Newton promised to redo his calculations and send them on to Halley by the London post.

Once again Halley's patience was sorely tested. Three months passed without a single word from Cambridge. What Halley did not know was that Newton had solved the problem of the elliptical orbit by employing a different mathematical method than before, but he was not satisfied. He spent much of those three months working on a nine-page manuscript titled *De Motu Corporum in Gyrum* (On the Motion of Revolving Bodies). Finally, in November 1684, nearly eleven months after Halley, Hooke, and Wren had taken part in the discussion that had started the quest, a copy of *De Motu* arrived at Halley's doorstep in London.

Halley was astounded, for in his hands were the mathematical seeds of a general science of dynamics—the study of the relationship between motion and the forces that affect it. Not wasting a moment, he headed north to Cambridge a second time. He must find out whether Newton would agree to have his paper set before the Royal Society and published for all the scientific world to see.

On December 10, Halley rose to address his fellow members of the Royal Society and its new president, Samuel Pepys. He gave an account of his most recent visit with Newton and of a "curious treatise," *De Motu*. Halley's report was duly recorded in the minutes, and he was urged to push Newton to publish his little work as soon as possible.

At first Newton may have thought of *De Motu* as an end in itself. But once his creative powers were loosed, there was no checking their momentum. "Now that I am upon this subject," he wrote Halley in January 1685, "I would gladly know the bottom of it before I publish my papers."[9] In his mind's eye *De Motu* would serve as the germ of his masterpiece, the greatest book of science ever written.

Thus began eighteen months of the most intense labor in the history of science. In April 1686 Newton presented and dedicated to the Royal Society the first third of his illustrious work. He titled it *Philosophiae Naturalis Principia Mathematica* (Mathematical Principles of Natural Philosophy), commonly known as the *Principia*. A month later the Fellows agreed that the society should pay the cost of the book's publication, but that decision was reversed two weeks later when it was reported that the treasury was empty for having funded a slow-selling history of fishes. The members turned to Halley, who agreed to finance publication out of his own pocket and to act as the difficult Newton's editor.

It was a fortunate choice. No sooner had Newton presented his first installment than Robert Hooke raised his all-too-familiar cry of theft. He claimed to have formulated the inverse square law six years earlier, which he had communicated in a letter to Newton. While it is true that Hooke had corrected a rare blunder Newton made regarding the path of a falling body, this was a far stretch from proving that the greater the distance between a planet and the sun the less intense the gravitational attraction between them. In addressing Hooke's claim that Newton had stolen from him, the eighteenth-century French scientist Alexis Claude Clairaut later observed, "what a distance there is between a truth that is glimpsed and a truth that is demonstrated."[10]

Newton was predictably outraged when word of Hooke's charge reached him in Cambridge. He immediately dashed off an angry letter to Halley in which he threatened to withhold the rest of the *Principia*. "Philosophy is such an impertinently litigous Lady," he wrote, "that a man had as good be engaged in lawsuits, as have to do with her. I found it so formerly, and now I am no sooner come near to her again, but she gives me warning."[11]

Tall, dark-eyed, soft of face and manner, Halley was a presence pleasing to almost everyone. Despite many frustrations, his treatment of Newton was unwavering, ever polite and respectful from their first meeting to their last letter, a relationship destined to endure forty more years. He moved swiftly to calm the troubled waters. All Hooke desired, Halley wrote Newton, was to be mentioned in the preface of the *Principia*. It would be a fine gesture on Newton's part, and one that would cost him nothing. A still infuriated Newton reacted by going over his manuscript and crossing out every reference to Hooke. With that the storm passed and Newton agreed to go ahead with publication.

The *Principia* was not an easy book to read in Newton's day, nor is it now. After it was published, Newton was passed on the street by a student who is said to have remarked, "There goes the man that writt a book that neither he nor anybody else understands."[12] The same would be said of Einstein when his papers on the theory of relativity were published some 250 years later.

The first of the *Principia*'s three books deals with problems of motion involving no friction or resistance. Book II is concerned with the motions of fluids and the effect of friction on the motions of solid bodies in fluids. Important though these books are, it is Book III, entitled *System of the World*, that most concerns us.

According to Newton's first law, "Every body continues in its state of rest, or of uniform motion in a straight line, unless it is compelled to change that state by forces impressed upon it." As we have seen, Galileo was actually the first to formulate this principle. Taking up where the Italian left off, Newton recast and incorporated it into his own system of mechanics, or the behavior of matter. If no outside force acts on a body, it will continue to move at a constant speed in the same direction. If left alone, a planet will orbit the sun eternally.

Yet the planets, as Newton had demonstrated mathematically, circle the sun, tracing out elliptical orbits. Why is it that they do not move off into space in a straight line, as would be expected according to his first law? This is where Newton's second law comes into play: "The change of motion [of a body] is proportional to the motive force impressed; and is made in the direction of the straight line in which that force is impressed." Stated in less rigorous terms, this law tells us that the orbiting planet is pulled at a right angle toward the sun. Its natural tendency to move outward into space, or what Christiaan Huygens called its "centrifugal" force, is perfectly balanced by the sun's inward pull, or what Newton termed "centripetal" force. One of the best ways to illustrate this principle is to whirl an object on a rope or a string over one's head. The object can be likened to a planet, the anchoring hand to the sun, while the string acts as the "force" that keeps the object from flying off into the blue.

But what of the string itself? No such visible tether binds a planet to the sun. Enter Newton's third law, which was uniquely his own: "To every action there is always opposed an equal reaction: or, the mutual action of two bodies upon each other are always equal, and directed to contrary parts." Hence, if one body acts upon another at a distance, the second also acts on the first with an equal and opposite force. The moon pulls the Earth with the same force with which the Earth pulls the moon. This is no less true of the Earth and an *apple*, except that in this instance the force exerted causes the apple to visibly change its position, while the Earth, because of its far greater size, seems totally unaffected. With these three laws of motion, Newton founded the branch of modern physics we call dynamics.

The genius of Newton's achievement becomes even clearer when we focus on the third law. Gravity acting at a distance could no longer be thought of as something peculiar to the sun and the planets; it applies to every object in existence, no matter how large or how small. As a universal property of all bodies, its force is dependent solely on the amount of matter each body contains. Or, as Newton announced in Proposition VII of Book III: "Every particle of matter attracts every other particle with a force proportional to the product of the masses and inversely proportional to the square of the distances between them." With this elegant principle, he had made the universe a "democracy" by treating all objects as equals. Everything—from the smallest of atoms to the largest of planets—obeys the same unchanging law, as profound a thought as has ever crossed a human mind.

Newton was ready now to demonstrate how the law of universal gravitation accounts for phenomena that had confounded the greatest minds in history for centuries. Or, in his words, "to admit no more causes of natural things than are both true and sufficient to explain their appearances."[13]

He focused his attention on Saturn's orbit of the sun, which he had attempted to calculate accurately for several years. Had it only been a matter of determining the mutual attraction of two bodies, the solution would have been relatively easy. But, as Newton well knew, the problem is complicated by the fact that Saturn's movement is influenced by other bodies as well, most notably its neighboring planet Jupiter. Although the sun, which contains a thousand times more matter than all the planets put together, is the solar system's dominant body, Jupiter is sufficiently large to produce small changes, or perturbations, in Saturn's orbit. Thus while Saturn follows an elliptical path around the sun, it wavers during the course of its journey, as though buffeted by an ill wind. Even Newton, armed as he was with the calculus, could not come up with anything more than a general solution to what is called "the three-bodied problem," one of the most difficult in physics. Indeed, he once remarked that trying to resolve it made his head ache, which he treated by tying a band of cloth around his head and twisting it with a stick until the reduced circulation dulled the pain. His pioneering study of planetary perturbation was refined over the decades until, in 1846, the planet Neptune was discovered by means of its gravitational pull on Uranus, the first such body to be discovered on the basis of theoretical calculations alone.

No less intriguing to Newton were the marked irregularities in the moon's orbit, a phenomenon that had puzzled astronomers for centuries. While the lunar orbit is dictated by the pull of Earth, it too is influenced by the enormous mass of the sun. And unlike the perturbations of the planets, those of the moon are both more numerous and more pronounced. By employing a complex system of calculations, Newton was able to account for most of these major disturbances. He did the same for Jupiter and its satellites, which Galileo had discovered in 1609 while gazing through his homemade telescope.

Among the more interesting deductions found in the *Principia* is Newton's assertion that the Earth and the other planets are oblate bodies. This means that they are flattened somewhat at their poles, or axes, and bulge to some degree at their equators. In terms of Earth, the equatorial bulge means the planet's surface is a few miles higher at its central belt than at the North and South Poles, a seemingly small difference but one fraught with significant implications.

Elsewhere in the *Principia*, Newton had demonstrated that a perfect sphere attracts another body as if its mass were concentrated at its center. However, oblate bodies, such as Earth, do not.

This means that the intensity of the planet's gravitational field will not be exactly the same everywhere. The Earth pulls and is in turn pulled by the moon with a slightly off-center attraction, the line of pull being strongest at the equatorial bulge, where matter is most concentrated. In effect we are dealing with a giant top slightly overloaded on one side. This causes the planet's axis to change its angle of rotation very slowly, so as to trace out the shape of a cone in the heavens. Astronomers call this "the precession of the equinoxes." It was first observed by Hipparchus, a Greek astronomer of the late second century B.C., but its explanation defied solution by the best of minds, including a deeply frustrated Copernicus. Newton undertook the calculation of this conical motion, which he accurately attributed to the slightly off-center attractions of the moon. He found that it takes 26,000 years for the Earth's axis to complete the cone. Once again genius had triumphed by explaining a mystifying phenomenon and calculating the time frame in which it occurs, a far cry from the simple suggestion that an apple falls to the ground because the Earth attracts it.

Of all the riddles that bedeviled astronomers, none had proved more baffling than the eternal rise and fall of the seas. Isaac Newton ended their frustration with a stroke of his pen: "That the flux and reflux of the sea arise from the actions of the sun and moon."[14] By applying the law of gravitation to the problem, Newton found that the power of attraction is greater on the waters facing the attracting body than on the Earth as a whole, and greater on the Earth as a whole than on the waters on the opposite side. Because of the moon's nearness to Earth (240,000 vs. 93,000,000 miles for the sun), the satellite's gravitational pull gives rise to the main tides. Its chief effect is to cause a pair of waves, or ocean humps, of tremendous area to travel around the Earth once in a lunar day, or a little less than twenty-five hours. The sun's attraction produces a similar but lower pair of waves that circle Earth once in a solar day of twenty-four hours. The effect of these two pairs of waves periodically overtaking each other accounts for the spring and neap tides. The spring tides, the more marked of the two, occur when the sun, moon, and Earth are in a line and exert maximum gravitational pull. The weaker neap tides result when the sun and moon pull at right angles to each other. While Newton's calculations were not refined enough to accurately predict the height of a tide at any given place in the world, another tremendous advance in scientific knowledge was his to claim.

As a young man, Newton had written of staying up all night to watch comets, and of becoming ill for lack of rest. Until shortly before his birth, these mysterious visitors had been regarded as nothing more than vaporous exhalations from Earth released into the higher regions of the atmosphere. More recently, comets were thought of as independent celestial bodies, but no one had been able to account for their seemingly irregular movements across the midnight sky.

Believing comets to be composed of solid matter, Newton reasoned that they must be subject to the same gravitational forces as the planets. Yet when he employed observational data collected in part by the astronomer royal, John Flamsteed, with whom he quarreled almost as violently as he did with Hooke, he found that the motions of comets were much more complex than those of the planetary bodies. As a result, he set out to prove that besides elliptical orbits, celestial bodies could move in paths tracing out sharper curves than previously imagined. He soon gave substance to this conjecture by determining that comets "move in some of the conic sections" about the sun. The natural philosopher then plotted the curve of the so-called Great Comet of 1681.

Based on what he read in the *Principia*, Halley was attracted to cometary theory as well. The gifted astronomer took particular interest in the orbit of another brilliant comet, which he had closely observed in 1682. Then, through a painstaking search of past records worthy of Newton himself, he found that there had been similar sightings in 1607 and 1531, or about once every seventy-five years. Could not these sightings, Halley thought to himself, indicate the periodic return of the same object? He calculated the orbit on the assumption that it was, and predicted its return in 1758, give or take a year. The streaming body, which now bears Halley's name, was next observed on Christmas Day, 1758 (the 116th anniversary of Newton's birth), by an amateur astronomer named George Palitsch. Like clockwork, Halley's comet has since shown up three more times, reaffirming Newton's reduction of yet another great mystery to mathematical rule.

When we step back and look at Newton's universe from afar, what is it, exactly, that we see? According to the *Principia*, we peer into a seemingly endless void of which only a tiny part is occupied by material bodies moving through the boundless and bottomless abyss. Newton's followers would liken it to a colossal machine, much like the clocks located on the faces of medieval buildings. All motions are reduced to mechanical laws, a universe where human beings and their world of the senses have no effect. Yet for all its lack of feeling, it is a realm of precise, harmonious, and rational principles. Mathematical laws bind each particle of matter to every other particle, barring the gate to disorder and chaos. By flinging gravity across the void, Isaac Newton united physics and astronomy in a single science of matter in motion, fulfilling the dreams of Pythagoras, Copernicus, Kepler, Galileo, and countless others in between. And while Newton was unable to discover the true cause of gravity itself, a giant riddle still, the laws he formulated provide convincing proof that we inhabit an orderly and knowable universe.

Eight

"Your Most Unfortunate Servant"

The book that supposedly nobody could understand was finally published in July 1687. Halley, who had put up his own money to see it through the press, set the price of the leather-bound volume at nine shillings, a bargain, considering the riches between its elegant covers. Not satisfied, he offered a second and cheaper version of the work in hopes that more people would read the revolutionary treatise. So awed was the astronomer that he paid its author yet another compliment. He wrote a poem titled "Ode to Newton" and inserted it at the beginning of the *Principia*. Its final line speaks volumes: "Nearer to the gods no mortal may approach."

Halley's sentiments were soon echoed by others. From Scotland the noted mathematics professor David Gregory wrote Newton that "you justly deserve the admiration of the best Geometers and Naturalists, in this and all succeeding ages."[1] A stunned French mathematician, the Marquis de l'Hôpital, asked

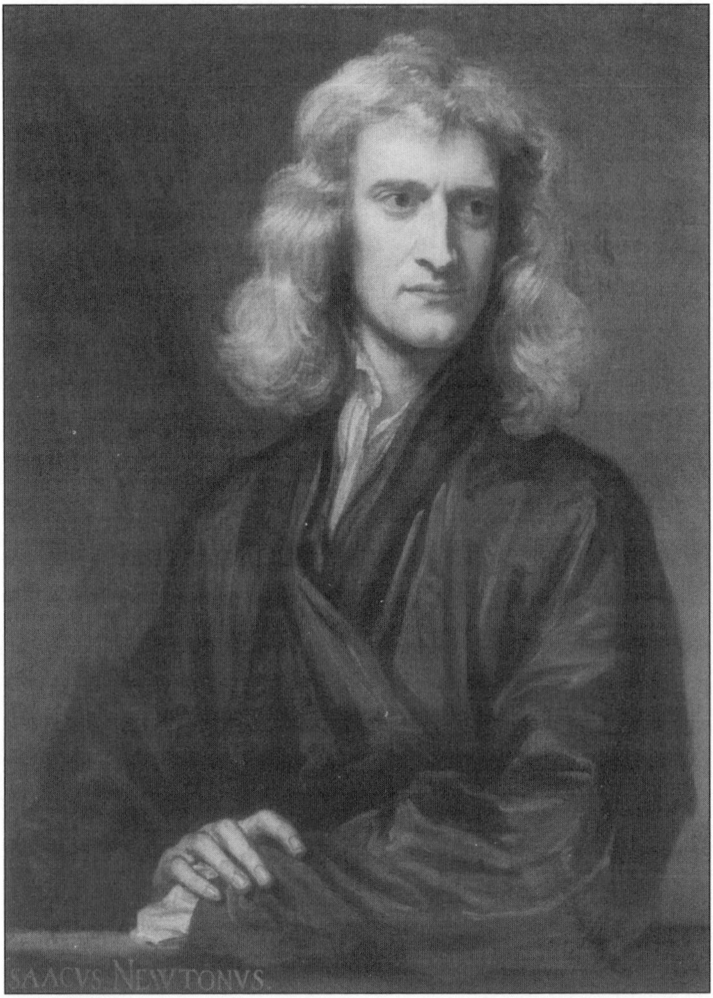

Sir Godfrey Kneller, the most popular portrait painter of his time, produced the first portrait of Newton in 1689, when Newton was forty-six. *Courtesy of the Trustees of the Portsmouth Estates*

his English friend Dr. John Arbuthnot, "Does Newton eat & drink and sleep? Is he like other men?"[2] Another young teacher of mathematics named Abraham de Moivre, who was destined to become a disciple of Newton, sought to master the *Principia* by stuffing his pockets with pages torn from the book, which he read while traveling from one pupil's home to another. Some of the highest praise came from Newton's small social circle, such as the aging Dr. Humphrey Babington, a fellow citizen of Lincolnshire and a Trinity College scholar. After wrestling with the scientific masterpiece for several weeks, Babington remarked that learned men "might study seven years before they understood anything of it."[3]

Despite his hard-earned fame, Newton continued his secretive ways. His normally fitful correspondence dwindled to less than a trickle. Only rarely did he emerge from the cocoon of his study to visit London or his Woolsthorpe property, where his tenants took advantage of his long absences by allowing the buildings and fences to fall into ruin. Rents were overdue by as much as three years, but the landlord was so well off that it little mattered.

Events took a sudden turn on January 15, 1689, when the Cambridge University senate met to elect two delegates to the so-called Convention Parliament. The previous year King James II, who hoped to restore Catholicism to England, was forced to flee to France during a political struggle known as the Glorious Revolution. Respected for his anti-Catholic views as well as his intellect, Newton gained a seat in the new Parliament. A week later he was in London, where he took his place in the House of Commons at daybreak.

Sensing the historic importance of the moment, Newton arranged to have his portrait done by Godfrey Kneller, the principal painter of the day. At age forty-six the subject is at the height of his mental powers and on the threshold of international fame. His silver hair is thick and flowing, the myopic eyes somewhat protuberant but piercing, the angular chin deeply cleft, the mouth sensuous and delicately formed. The long, slender fingers of Newton's right hand, which are visible from beneath his academic gown, seem more like those of a performer of music than the choreographer of matter's deterministic dance. Kneller's midlife portrait is a foreshadowing of that famous Jovian look borne by the aged Newton, who shared with Einstein the aspect of one present at the Creation.

Newton voted with the parliamentary majority in declaring the English throne vacant. Within days Prince William of Orange, a province of the Netherlands, was crowned King William III of England. William's wife, Mary, the Protestant daughter of the exiled James II, was crowned queen. The monarchs paraded through the streets of London to the cheers of the multitude, while Newton and his fellow members of Parliament marched respectfully behind. In the weeks that followed, Parliament laid the basis of a constitutional monarchy, limiting the power of the throne and granting religious freedom to all Christians, except Catholics and those who opposed the doctrine of the Trinity, a matter on which Newton wisely remained silent. Most important, Parliament enacted the Bill of Rights, which would become a model for the first ten amendments to the United States Constitution. Among its provisions was the right of Englishmen to petition their government for a redress of grievances, freedom from excessive fines and bail, a prohibition against cruel punishments, and the guarantee that taxes could be raised only with the consent of Parliament. The long-standing belief that the king was chosen by God and thus responsible to God alone had been dealt a deathblow.

Having personally reduced the motion of matter to a single universal law, Newton participated in the most important session of Parliament in English history. The branches of government, like the planets, were being balanced for the first time. Still the records of these deliberations contain not a single reference to him. According to a frequently repeated story, he spoke but once, and that was to ask an usher to close a window because of a chilling draft.

Yet the year Newton passed in London had seen a marked change in the man. He spent time in the company of his fellow natural philosophers at the Royal Society and made a number of influential friends. Among these was the political philosopher John Locke, whose writings on constitutional government had won him the reputation as a champion of liberty. The parliamentary leader Charles Montague, better known as Lord Halifax, also befriended Newton, as did the visiting Dutch astronomer Christiaan Huygens, who apparently bore no grudge over the harsh criticism Newton leveled at him in 1673 for disagreeing with his theory of light.

Indeed, Newton and Huygens became involved in a minor scheme to further Newton's interests. The provost of King's College at Cambridge was dying, and Newton's friends persuaded him to apply for the post. Accompanied by Huygens, he met with the new king to present his case. The royal audience went well, and it appeared that Newton would be awarded the position, never again having to speak into a dim and empty lecture hall. Then it was discovered that the incoming provost of King's must have taken Holy Orders and be a fellow of the college. Because Newton met neither of these qualifications, he was denied the position. He returned to Trinity College in February 1690 a disappointed man, but one with high connections and a new appreciation for the world beyond his quiet university town.

A brisk correspondence between Newton and Locke was launched in the months that followed and continued until the latter's death in 1704. Proof that Newton had his eye on another position is contained in his query to Locke regarding the Earl of Monmouth, an influential supporter of the new government and king. Locke had asked Monmouth's assistance in finding his friend a public post, or preferment, as it was commonly known. Unfortunately for Newton, the political winds had shifted in London and nothing could be done, at least for the moment. A second scheme involved Locke's intimate friends Sir Francis and Lady Masham. Newton and Locke visited Oates, the Mashams' estate in Essex, to set forth Newton's case to become comptroller of the mint. This strategy also came to nothing, leaving Newton embarrassed and bitter.

Newton and Locke had been joined at Essex by a third person, the young Swiss mathematician Nicolas Fatio de Duillier. Newton had met Fatio in London, and their friendship soon developed into the most emotional attachment of Newton's adult life. Fatio was no less taken by the author of the *Principia*. Whereas Halley had likened Newton to the gods, Fatio believed him to be one.

If one is to judge by his only portrait, Fatio was far from a handsome man. His face is dominated by a great Roman nose, made even more prominent by a high, sweeping forehead, small mouth, and rather sharply pointed chin. Only his eyes are compelling: keen and intelligent, they seem capable of transfixing anyone.

Fatio wrote his old philosophy teacher back in Geneva that Newton was both the greatest mathematician ever to have lived, as well as the most worthy gentleman. He hoped to become a citizen of England and reside with the natural philosopher from whom he could learn "pure truth." If only he had the money he would erect a monument to his friend, a testament to future generations that at least one Newton admirer recognized his true worth.

Newton and Fatio met several times during the next few years. A typical visit had Newton taking the coach to London, his mathematical and religious papers in hand, where an anxious Fatio awaited his arrival at a comfortable inn. Newton gave his young friend access to his most private writings, including those which, if published, would have seen him declared a heretic. Such trust was rare for Newton. The only others to gain access to his innermost thoughts were Locke and a small number of young disciples, such as the Cambridge mathematician William Whiston.

Before long Fatio began to regard Newton as his private possession. He tested the bond between them by faking illness and then asking special favors of his friend, which an anguished Newton willingly granted. Worse yet, Fatio confused Newton's genius with his own mental skills. He would later claim, much to the amusement of anyone who would listen, that he had found the cause of gravity and that Newton had confirmed it. This was nonsense pure and simple, and it may have contributed to the eventual severing of the friendship.

It was against this emotionally charged background that Newton decided to have one last try at alchemy. In the spring and summer of 1693 he completed five chapters of a manuscript on the multiplication of metals. For a brief while the secret of the ages seemed within his grasp. He watched transfixed as gold mixed with a special mercury swelled on being exposed to fire. But he soon discovered that something was wrong. The amount of gold supposedly feeding off the mercury had not increased one bit.

Newton crossed out several passages of his projected treatise and tried again. And again he failed. Finally, after working at the ragged edge for months, the work was abandoned forever.

In September 1693 Samuel Pepys, who had served as president of the Royal Society while Halley was seeing the *Principia* through the press, received a distressing letter. "I am extremely troubled at the embroilment I am in," Newton wrote, "and have neither ate nor slept well this twelve month, nor have my former consistency of mind. I never designed to get anything by your interest, nor by King James's favor, but am now sensible that I must withdraw from your acquaintance, and see neither you nor the rest of my friends any more."[4]

Pepys was not only deeply shaken but innocent of Newton's accusations. By this time James II had been in exile for years and Pepys wielded no influence at the court of William III. Three days later, John Locke received an even more disturbing and vengeful letter from his Cambridge friend. Newton accused him of attempting to "embroil me with women." When he had been informed that Locke was ill, Newton replied, "twere better if you were dead." He had also taken Locke for an "atheist" and charged him, as he had Pepys, with unscrupulous dealings in his attempt to gain preferment. Newton signed himself "your most humble & most unfortunate Servant."[5]

The only good thing about the letter was its conclusion, in which Newton begged his friend's forgiveness. A deeply concerned Locke wrote back immediately: "Give me leave to assure you that I am more ready to forgive you than you can be to desire it."[6] Newton, who by now was feeling much better, attempted to explain himself in a second letter. The previous winter he had gotten into an "ill habit" of sleeping by the fire. His health suffered and he then caught "a distemper which this summer has been epidemical [and] put me further out of order." When he first wrote Locke he "had not slept an hour a night for a fortnight together and for 5 nights together not a wink."[7]

Just to be certain that Newton was his old self again, Pepys and Locke asked their mutual friend John Millington, a Fellow of Cambridge's Magdalene College, to pay a call on Newton. Millington was happy to report that Newton seemed normal, if a bit depressed. During this visit, Newton confessed to writing Pepys "a very odd letter" for which he was "very much ashamed." As in his recent letter to Locke, he blamed his temporary derangement on a prolonged lack of sleep and a bout of distemper, a general term then used to describe many kinds of illnesses.

Few scholars have taken Newton at his word, and various theories have been advanced to explain this dark passage. In fact, Newton had been through what may have been a similar, though less severe, episode many years before. As a student, he had stayed up all night watching comets and became ill for lack of sleep.

But that is hardly the whole story. One theory holds that Newton took leave of his senses because many of his irreplaceable papers were destroyed in a fire. The supposed culprit was his pet dog named Diamond, who overturned a candle in Newton's chambers, causing his anguished master to exclaim, "O Diamond! Diamond! thou little knowest the mischief done!"[8] The problem is that Newton never owned a dog at Cambridge, or for that matter a cat. And had there been a destructive fire, word of it would have almost surely gotten around.

A more plausible explanation is the much discussed theory of poisoning by lead and mercury, two metals commonly found in Newton's crucibles. The trembling Mad Hatter of *Alice in*

Wonderland fame was a victim of lead poisoning, as were his fellow hatmakers who used considerable amounts of the metal in their trade. Besides uncontrollable tremors, regular exposure to lead and mercury causes loosening of the teeth, darkening of the nails, weight loss, jaundice, sluggishness, and early aging. As it happened, Newton displayed none of these telltale symptoms. His handwriting from the period is as steady as a rock; though he lived to be eighty-four, he lost only a single permanent tooth; he loved sweets and gained weight as he got older; and he aged more slowly than many of his friends. Most important of all, once heavy metals enter the body they remain for the rest of the victim's life. Though Newton breathed in copious amounts of them, his constitution was strong enough to withstand their presence without showing any symptoms. Otherwise, he would have taken ill again and again, becoming increasingly debilitated with time.

It is more than likely that Newton's illness resulted from a combination of several factors, both physical and mental. Humphrey Newton vividly described how his employer scarcely slept while undertaking alchemical experiments. In 1693 Newton was working at a fever pitch. Moreover, after some initial success, his experiments were yielding nothing but frustration. To one who had already unlocked many of nature's greatest secrets, this was both a depressing and fearsome thought—an Isaac Newton who had reached the limit of his powers.

Moreover, Newton, as he informed Locke, had been ailing on and off for a year. It seems more than a coincidence that his friend Fatio also claimed to be ill during this same period. Then, for some unknown reason, Newton suddenly broke off their friendship and would have little to do with the man for as long as he lived. This, too, occurred about the time he wrote his pathetic letters to Pepys and Locke. Thus Newton's mind was still intact, but his spirit was no longer willing. He who had attempted to rise above human art and industry now longed to leave his beloved Cambridge, the sooner the better.

Nine

Mark of the Lion

In September 1695 Isaac Newton suddenly disappeared from Cambridge without explanation. He returned two weeks later but told no one where he had been. Rumors began to circulate that he secretly had gone to London, and finally secured the government post that his friend John Locke had been unable to deliver. These rumors were given substance in late November when the mathematician John Wallis wrote Edmond Halley from Oxford: "We are told here Newton is made Master of the Mint: which, if so, I do congratulate him."[1] Halley, who had kept in touch with Newton, was aware that negotiations were under way, and that Newton's cause was being championed by his ally Charles Montague, a powerful political figure who had the king's ear. Halley also knew that Wallis was somewhat misinformed. Newton was not a candidate for the position of master of the mint but rather that of warden, or second in command. In March 1696, Montague again summoned Newton to London. If all went well, the appointment would be his together with the handsome salary of five or six hundred pounds per year. Without even pausing to draft a reply, Newton boarded the coach for the capital and his long-awaited audience with King William III.

No account of the meeting survives, doubtless because there was nothing special about the occasion—except of course to Isaac Newton, the son of an illiterate yeoman. The warrant confirming his appointment was drafted two days later. He quickly returned to Cambridge and set about packing trunks with the accumulation of thirty-five years' hard labor, including thousands of pages of manuscripts containing millions of words of correspondence, mathematics, optics, alchemy, and religious writings. Such was his haste that he left many personal possessions behind, including his furniture and alchemical equipment. These things were never reclaimed and comprised something of a Newton museum. His rooms were proudly shown to visitors long after his death with the boast that every relic was preserved down to the smallest detail.

As a result of the Great Fire of 1666, the London of merry old England was no more. Upon its ashes had arisen a teeming, overcrowded city of some 750,000 souls, second in population only to Paris among Europe's major urban centers. Stretching from Tower Hill to the Houses of Parliament in Westminster, the city grew up along a great boulevard connected by London Bridge, the only span across the river Thames. This sprawling waterway, which drains into the North Sea, ebbs and flows with the tides, and was forever crowded with hundreds of ships and countless smaller craft, the latter providing a speedier and safer means of transportation than the teeming city streets. But in late autumn and winter, piercing mists and thick fogs rolled off the river's foul waters, mixed with the hovering smoke from countless chimneys, and filled the streets with a brown, poisonous cloud that drove all but the foolhardy and desperate indoors.

Much of London's work was done by a rough-and-tumble population of fiercely independent cockneys who inhabited the East End—watermen, porters, dockmen, and day laborers. Above them was a middle class of shopkeepers and artisans. At the top of the social order stood the wealthy merchants, bankers, and political officials, who resided in the best neighborhoods where squalor and congestion were kept at arm's length.

Beyond the old city walls first erected by the Romans, who named their frontier outpost Londinium, dwelt the huge underclass comprised of the poor, the homeless, and the criminal, hopeless men and women with whom Newton would soon have reason to deal on a regular basis. By the tens of thousands they occupied unsanitary shacks facing dark and dangerous passages choked with garbage and human waste pitched from glassless windows above. Here, in the so-called Liberties, violence walked hand in hand with the crudest of pleasures, while the powers of the lord mayor and the constables counted for little. Public floggings seldom discouraged illegal conduct, nor did the regular executions, despite the fact that they drew vast, jostling crowds on "Hanging Day." Those condemned to die were borne amid jeering throngs from Newgate Prison to Tyburn, the place of execution at one corner of Hyde Park. The wealthy paid handsomely for choice seats on the wooden galleries surrounding the gallows; the remainder of the crowd pressed forward in hopes of gaining a clearer view of the spectacle. Those convicted of state crimes such as treason were beheaded by John "Jack" Ketch, the infamous executioner of the day. Death on the gallows was the penalty for forgers, and it fell to the warden of the mint to supply the evidence needed to convict those who counterfeited the king's coin.

The royal mint was located in one of England's most forbidding and historic structures—the great Tower of London—a massive limestone fortress situated atop a hill on the north bank of the Thames. The tower's soaring walls were guarded by a moat, which was still flooded in Newton's day. Once behind this outer line of defense, the visitor encounters a second tower-ringed wall, but of even greater height than the first. Inside the second wall are the soldiers' barracks, storehouses for powder and arms, and other buildings dating from the Middle Ages. At the center of it all stands the keep, or dungeon, known as White Tower. Constructed in the time of William the Conqueror, who successfully invaded England in 1066, White Tower once served as the court of Plantagenet kings born and nurtured within its mighty walls.

Through the centuries, the Tower of London was home to England's most famous prisoners. The future Queen Elizabeth I, daughter of Henry VIII, had entered through Traitor's Gate while still a princess and wept against the tower's stone walls before regaining her freedom. Certain of her father's luckless wives and political servants had suffered far worse. Ann Boleyn, Catherine Howard, Thomas Cromwell, and Sir Thomas More all breathed their last on Tower Green after enduring miserable confinements. Nor, after she became queen, did Elizabeth herself shrink from her duty as she saw it. Lord Essex, once among her court favorites, met his God on Tower Green at the hands of the axman, while the swashbuckling explorer Sir Walter Raleigh, who had supplied her majesty with captured treasure galore, spent three long imprisonments in the tower on her orders.

Like all employees of the mint, Newton took the oath of secrecy, pledging not to reveal how new coins were made in hopes of keeping forgers in the dark. By 1695 counterfeiting had become so widespread that every English coin in circulation was to be replaced by a freshly minted one. This monumental task required mint employees to work double shifts and led to the opening of branch mints in other parts of the country. Under the new warden's watchful eye, rivers of gold and silver coins began pouring forth from the lumbering stamping machines. And with his superior Thomas Neale frequently absent, Newton quickly became master of the mint in everything but name.

Still he could not savor his success until he had dealt with a colorful rogue whose rare but misused gifts nearly proved a match for his own. William Chaloner was a natural-born swindler and

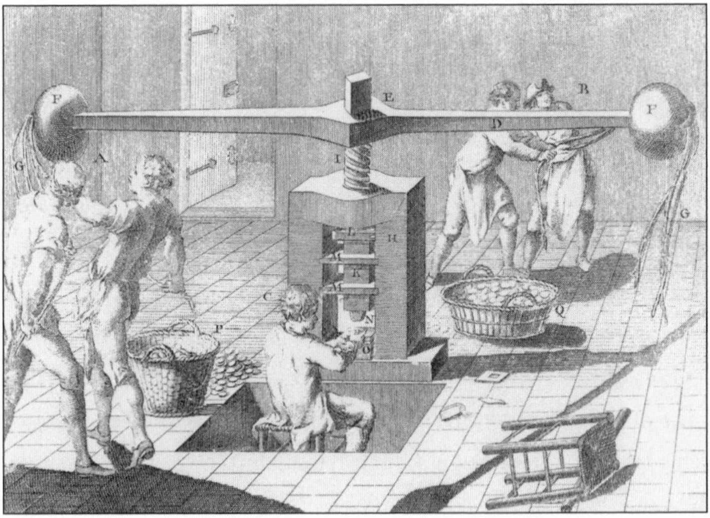

A coining press in operation. When he began work at the mint, Newton took an oath of secrecy, vowing not to reveal how new coins were made. *Library of Congress*

cunning forger who first made his way in London by fashioning worthless tin watches, which he hawked in the streets for whatever the gullible would pay. He soon formed a partnership with another rogue, and together they "set up for Piss-Pot Prophets, or Quack-Doctors," according to an anonymous pamphlet of the times.[2] Chaloner next bribed a craftsman to teach him the secret of enameling wood in the Japanese manner. Applying this knowledge to the study of metals, he was soon making counterfeit coins that easily passed in the marketplace for the real thing.

Meanwhile, Newton had quickly familiarized himself with the criminal element of London's vast underworld. His many contacts, any one of whom might have done him in during an unguarded moment, included criminals of every stripe—murderers, thieves, beggars, swindlers, and muggers. He learned of bribes to destroy evidence, of hideaways in outlying villages, of secret meetings in taverns and dark attics. He developed contacts with men and women of many assumed names, hired undercover agents, and occasionally donned disguises of his own. Those who refused to cooperate in his investigations were brought before him in chains for questioning in the tower.

Believing himself to be in the service of a higher morality, Newton wanted nothing more than to arrest and convict Chaloner for his many crimes. But he was no ordinary criminal of the type Newton had grown used to dealing with. Though he was rich and rode about London in an elegant carriage with a beautiful mistress at his side, the master thief was less interested in wealth and the false respectability it purchased than in the playing of the game. He loved the fraud he called "funning," the plotting, the trickery, the sleight of hand that matched his wits against the talent and the resources of the government. When confronted with evidence linking him to a serious crime, he repeatedly wriggled free by resorting to the big lie, a device he jokingly referred to as "bubbling." By the time he crossed swords with the warden of the mint, Chaloner was nothing short of effervescent.

The forger made no greater mistake than when he boldly appeared before a committee of Parliament to recommend improvements in the minting of new coin. Newton was outraged and let it be known that Chaloner was his number one target. Chaloner responded by taunting "that old dogg" Newton, who, in a fit of anger, vowed to hang the villain. It was at this point that Newton decided to stretch the law in the interest of ridding the world of his clever tormentor.

Without warning, Chaloner was arrested and clapped in irons behind the dank walls of Newgate Prison. In return for certain favors, Newton employed three convicts to gain Chaloner's confidence and pass on his tales of criminal deeds. The ending, which had been so slow in coming, lasted only three weeks. Chaloner was placed on trial for high treason and found guilty, based largely on the testimony of Newton's questionable witnesses, all of whom were confessed criminals. Chaloner's protests that he had been framed fell on deaf ears, and the convicted man's date with the hangman was set.

Ironically, Chaloner's only hope of mercy rested with Newton. Shortly before his scheduled execution, he wrote an emotional letter to the warden: "I am going to be murdered although perhaps you may think not but tis true I shall be murdered by the worst of all murders that is in the face of Justice unless I am rescued by your merciful hands."[3] Days later Chaloner got his answer when he was loaded onto a cart and paraded two miles through the streets to the gallows in Hyde Park. There he repeated his claim, crying out hysterically to the assembled crowd that he was being murdered by lies and injustice. Few took notice, for it was a rare criminal indeed who did not proclaim his innocence when faced with the moment of death. Newton himself was not present at the hanging, but the man whom Chaloner had compared to an old dog had the last word. "Criminals, like dogs," he remarked, "always return to their vomit."[4]

Even as Newton pursued criminals and raised the production of coins to a record 100,000 pounds per week, his reputation as a natural philosopher continued to grow. In January 1697 he received a letter from Johann Bernoulli, a distinguished mathematics professor at the University of Basel in Switzerland. The letter contained two difficult mathematical problems, one of which had been published in a scientific journal six months earlier but remained unsolved. The other had been solved by the gifted German philosopher and mathematician Gottfried Wilhelm von Leibniz, but Bernoulli had decided to keep Leibniz's solution a secret in order to test Newton's powers.

Newton received Bernoulli's letter late one afternoon, after a long and tiring day at the mint, which was in the middle of the recoinage. The first of the challenge problems required Newton to determine the curve in which a heavy body falling under the force of its own weight will descend most rapidly from a given point to another given point. The second problem was even more complex and required a series of steps that had so far baffled all seekers save Leibniz.

Forgetting his dinner, Newton immediately set to work. Some twelve hours later, at four in the morning, he had the correct answers, yet still refused to go to bed. He drafted a letter to the president of the Royal Society and included the solutions, which, according to his instructions, were to be published anonymously in the *Philosophical Transactions*. When the journal came off the press a few weeks later, mathematicians were stunned. Among them was Johann Bernoulli, who wrote a friend that only Isaac Newton could have succeeded where others had failed: "I can tell the lion by the mark of his claw."[5]

Having met and been honored by the king, Newton was the subject of royal attention from another quarter. In June 1697 Peter the Great, the young czar of all the Russias, left his homeland on a visit to England and Western Europe. Standing six feet seven inches, a height that would have made his skeleton the prize of every scientific society in existence, Peter carried a seal with the inscription: "I am a pupil and need to be taught." Everything involving the new science and technology interested him. Indeed, the czar planned to take back much of what he learned to his homeland, which had barely emerged from the Middle Ages. After visiting England's shipyards, its main cannon factory, and the Royal Observatory at Greenwich, he specifically requested a meeting with the author of the *Principia*. It took place in the Tower of London, where Newton talked science with the czar and gave him a tour of the mint, whose great hammer presses, driven by ropes and human muscle, struck coins on different beats, like giant cuckoo clocks gone mad.

The czar and his retinue settled in at Sayes Court, which had been completely redecorated in honor of the Russians' visit. The first clue that something was seriously amiss came in the form of a letter from one of the English servants: "There is a house full of people, and right nasty."[6] Not until Peter and his court departed did the full extent of their wildly destructive behavior become evident. The carpets were so befouled that the floors beneath them had to be torn up and replaced. The windows in nearly every room were smashed, causing the occupants to reduce some fifty elegant chairs to kindling in an effort to stave off the cold. Featherbeds were shredded "as if by wild animals," and twenty paintings were riddled with holes from frequent sessions of indoor target practice. Neighbors observed the Russians playing a game with wheelbarrows, an invention unknown back home. They took turns—a revoltingly dirty Peter included—running one another through the magnificent four-hundred-foot-long holly hedges, leaving gaping holes in their wake.

Shortly after Peter left England, Newton was visited by Jaques Cassini, son of the noted astronomer Giovanni Domenico Cassini. As an agent of Louis XIV, the famous Sun King who had established his French court at Versailles surrounded by orange blossoms and a thousand compliant nobles, Cassini offered Newton a large pension. While Newton would never have said so in public, his religious views were such that he wanted nothing to do with a Catholic king of a Catholic kingdom, and so he politely refused the offer. In return, a rarely snubbed Louis excluded Newton's name from his list of nominees to the position of foreign associate of the French Academy of Science. But one year later, in 1699, the academy itself made Newton a foreign associate along with Johann Bernoulli, whose mathematical challenge he had so brilliantly met.

Except for an occasional visit to Lincolnshire, Newton's contacts with his family had been limited since the death of his mother. At Cambridge he had been content with the slow passage of the days and seasons, and with the isolation that allowed him to pursue his work undisturbed. Whether he liked it or not, London and the mint had transformed him into something of a man of action, requiring him to keep a household where important visitors were entertained over dinner and fine wine.

Never having married, Newton needed someone to look after his domestic needs and to sit at his table when he had guests. Rather than rely on a housekeeper, who might lack the required social graces, he turned to his niece Catherine Barton, the daughter of his stepsister Hannah. Catherine probably arrived at her uncle's house on Jermyn Street in 1696, when she was seventeen. By all accounts the young woman was strikingly beautiful. When she came down with smallpox, Newton sent her into the country to recuperate and worried that her lovely face would be permanently scarred: "Pray let me know," he wrote, "how your face is and if your fever be going. Perhaps warm milk from ye Cow may help to abate it."[7] Newton breathed a sigh of relief when he next set eyes on his niece, who was as beautiful as ever.

It was not Catherine's looks alone that charmed the usually stern Newton and the members of his social circle. She appears to have been the one relative of her uncle's who was gifted beyond the ordinary. Not only was Catherine graceful and pleasant to gaze at, but she was witty and well read, traits that gained her many admirers among important political and literary figures.

Newton's friend and patron Charles Montague wrote poetry in Catherine's honor and toasted her charms at the Kit-Kat Club, a fashionable association founded by the Whig politician Christopher "Kit" Cate. Jonathan Swift, the famed author of *Gulliver's Travels*, became her cherished friend. Swift thought Catherine an excellent conversationalist and a clever gossip whose company he never tired of. "I went to see [Miss] Barton this morning," he wrote a friend. "I love her better than anybody here, and see her seldomer. So it often happens in the world."[8]

When Swift spoke of love, he meant the deepest kind of friendship, but the hearts of others spoke romance. Rémond de Monmort, a French official, fell head over heels for Catherine during a visit to London, in spite of the fact that he had left a Madame de Monmort back in Paris. "I have adored her not only for her great beauty but for her lively and refined wit," Monmort confided to his friend Brooke Taylor of the Royal Society. "If I had the good fortune to be near her, I would henceforth become as awkward as I was the first time we met."[9] As far as Newton was concerned, no one since Fatio de Dullier had so deeply stirred his feelings. Catherine's gaiety, charm, and warmth made coming home from the mint on chill winter afternoons a genuine pleasure for the aging lion.

From Newton's point of view, Thomas Neale, the master of the mint, was nothing but a drone who performed no work and produced no honey. In other words, Neale was typical of most government officials who had secured comfortable positions through their high connections, then proceeded to enjoy the good life while underlings carried on the business of the day. The only good thing about Neale was that he gave the inexhaustible Newton free reign, enabling his second in command to master every detail of managing the mint. Then, on December 23, 1699, Thomas Neale died. Because the master had been seriously ill for months, it is likely that an agreement had already been reached concerning his successor. Two days later, on Christmas Day, Isaac Newton became master of the mint, a most welcome present on this, his fifty-seventh birthday. Whereas Neale had earned thousands of pounds every year, Newton had been working for a few hundred. This, too, suddenly changed. The new master saw his annual income multiply as much as tenfold, elevating an already wealthy man to the equivalent of a modern millionaire.

Ever conservative in matters involving his finances, Newton had not resigned the Lucasian professorship when he left Cambridge in 1696. Thus, if London did not suit his tastes, or if he suddenly found himself in danger of losing his position, he could always return to a post that had served him well. He cautiously waited another two years, during which he earned several thousand pounds as master, before he decided to cut his academic ties with the university. In December 1701, Newton gave up his professorship, but not before he handpicked his successor, a young mathematician named William Whiston who shared his secret and politically dangerous religious views.

As a gesture of respect, Newton was once again elected to Parliament by the university, and once again the records indicate that he said nothing during the long and often bitterly charged debates. However, the life of this Parliament was brief. A sickly William III breathed his last on May 7, 1702, and Princess Anne ascended the English throne. As was customary, the queen dissolved Parliament and called for new elections. When Newton was asked by the Cambridge authorities if he wished to run again, he politely declined: "And now I have served you in this Parliament, other gentlemen may expect their turn in the next."[10] Though he would remain a political appointee the rest of his life, Newton's days as a politician were over.

Ten

The Royal Society

Ever since he arrived in London, Newton had kept his distance from the Royal Society, the scientific institution through which he gained lasting fame. When asked why he did not become more involved, he claimed that his duties at the mint demanded all his time. There were those who knew better, however. The real reason Newton did not attend the meetings was an old grudge he had nursed ever since his reflecting telescope was carried to London by his friend Isaac Barrow.

It was then that Robert Hooke, the curator of experiments for the Royal Society, had argued that he had developed a smaller and more accurate instrument. Hooke had also criticized Newton's theory of light and claimed gravity as another of his mental inventions, causing a skeptical Edmond Halley to take the coach to Cambridge in hopes of settling the matter once and for all. And though they had engaged in an outwardly civil correspondence over the years, Newton hated Hooke, while Hooke wrote in his diary that he dreamed Newton had died.

At age sixty-eight, Robert Hooke, who was largely responsible for the breathtaking reconstruction of London after the Great Fire, had been reduced to a shriveled, pain-racked husk. He no longer attended meetings of the Royal Society and was living what one biographer termed "a dying life." Nearly blind and bedridden, the man of a thousand ideas—some brilliant and some crackpot—died on March 3, 1703. Hooke left no will and his money, which he liked to keep handy, was found locked in an iron chest.

Six months later, the Fellows of the Royal Society assembled for the annual election of its twenty-one-member council and officers. When the ballots were counted, Isaac Newton, who had suddenly taken a renewed interest in the society's affairs, was chosen president.

Newton may well have asked himself, "President of what?" The treasury of the society was bare; new scientific ideas were as scarce as hen's teeth; and the number of Fellows had declined from two hundred in 1680 to a little over one hundred on the eve of his presidency. Worse yet, few members attended the society's weekly meetings and fewer still the council, which sometimes made its decisions at gatherings little larger than the foursome needed for a game of bridge.

When discussions took place, they were mainly devoted to medicine or the anatomy of exotic beasts. The Fellows, many of whom were physicians, doted on live specimens of crocodiles, armadillos, and opossums. They were fascinated by presentations on the merits of various poisons employed by convicted murderers, and debated the medicinal value of pig's urine, a far cry from the days when the laws governing the universe were the main topic of concern. Matters had gotten so far out of hand that the Fellows became the butt of jibes by anonymous pamphleteers. They were accused of searching for a deer in the belly of a snake, of finding an egg in the tail of a barnacle, of sorting "lady's bugs," and of witnessing a shower of fish followed by a shower of butter in which to cook them. Indeed, it seemed that everyone was laughing—everyone, that is, but the new president, who was little known for his sense of humor.

So few members showed up at one of the first meetings chaired by Newton that it had to be canceled. The angry president reacted by drafting a master plan entitled a "Scheme for establishing the Royal Society," a blunt way of stating that the organization no longer existed and had to begin anew. First, he wrote, the society must return to its original mission, that of "discovering the frame & operations of Nature, reducing them (as far as may be) to general Rules or Laws, establishing those Rules by observations & experiments, and then deducing the causes & effects of things."[1] To those familiar with the new natural philosophy, Newton was calling for the application of Galileo's scientific method.

To carry out this work, Newton recommended the appointment of four paid "demonstrators" who would attend all meetings and present regularly scheduled experiments and lectures in mechanics and mathematics, optics and astronomy, zoology, chemistry, and botany. While he could never bring himself to mention Hooke by name, there is no question that Newton was thinking of just such a model multiplied by four. To accomplish these things the Royal Society had to have a permanent residence, dozens of active members, and a replenished treasury—matters that Newton addressed in the second part of his ambitious scheme.

Ever since receiving its charter in 1662, the society had met at Gresham College in Bishopsgate Street. This vast timber-and-brick mansion was originally constructed as a private residence by the wealthy merchant Sir Thomas Gresham, who toasted Queen Elizabeth with wine laced with the essence of pearls in the great dining hall. Having no rooms of their own, the Fellows assembled in the apartments of various professors over the years, the last being the quarters of the recently deceased Robert Hooke.

It was Newton's plan either to construct or to purchase a home for the society. To accomplish this he would have to raise funds from the Fellows, many of whose membership fees had not been paid for years. After much debate, the council approved the president's proposal that every candidate for membership must pay the admission fee and sign a bond for his weekly dues before his induction. Fellows who had not previously signed a bond were now called on to do so, and no member who failed to comply or who was behind in his dues could serve on the council. Despite many complaints and much grumbling, money began to trickle in. The society was also given some stock in the East India and the East Africa Companies, which had been formed to increase trade in the expanding British Empire.

Working closely and in secret with Royal Society Secretary Hans Sloane, Newton spent the next few years scouring London for a suitable headquarters while the treasury grew. Although regular meetings were suspended during the summer months, the council was suddenly called into emergency session in September 1710. With the help of Sir Christopher Wren, Newton had located a house in Crane Court off bustling Fleet Street, which was available for the sum of 1,450 pounds. Newton gained the council's approval, and when the society next met on October

26, the president proudly reported that the sale had been completed. Those on the council who had opposed his plan were soundly defeated when they stood for reelection a month later, a lesson to anyone who thought of crossing swords with the unforgiving president.

Ruling with the hand of a monarch, Newton had moved the Royal Society onto firm financial ground after fifty years of hand-to-mouth existence. When the planned remodeling of the new quarters was completed ten months later, the society assembled around its unchallenged leader like obedient members of court. According to the newly passed Orders of the Council, only the president was permitted to sit at the head of the table, with the two secretaries at the lower end, one on either side. No talking was permitted among members during the meetings unless they first addressed themselves to the president. Still not satisfied, Newton established the practice that the mace, the symbol of authority, could be placed on the table only when the president occupied the chair. On those occasions when he was absent and the vice president took his place, the mace remained in its stand. Dr. William Stukeley, the young Lincolnshire physician who was later befriended by Newton, described the scene: "Whilst Newton presided in the Royal Society, he executed that office with singular prudence, with grace and dignity. He was very careful of giving any sort of discouragement to all attempts of improvement in natural knowledge. There was no whispering, talking, nor loud laughters. Everything was transacted with great attention and solemnity and decency. Indeed his presence created a natural awe in the assembly."[2]

Fully assured of his place in history, Newton had his image reproduced ever more frequently as he grew older. He posed for a second portrait by Godfrey Kneller in 1702 and another by Charles Jervas only a year later, as he was about to become president of the Royal Society. In Jervas's work the natural philosopher sits stiffly in a high-back armchair, his head crowned by a flowing wig, his right forefinger pointing to an untitled book lying on a nearby table. The bulging eyes challenge the viewer to blink first, a contest Newton always wins. This is an image of complete control and supreme accomplishment, the very one he wished to project for future generations. He later presented the Jervas portrait to the Royal Society, where it is still displayed like a sacred object. Meanwhile, the only known painting of Hooke mysteriously vanished during the Royal Society's move from Gresham College to Crane Court, raising strong suspicions that Newton himself was responsible for its disappearance.

In April 1705, Queen Anne, accompanied by her simple but amiable husband, Prince George of Denmark, set out from London for the royal residence at Newmarket. The main reason for the journey was to attend the annual races, where the best horses in the kingdom charged across Newmarket heath and over the Devil's Dyke, an ancient earthwork. The party then made its way to nearby Cambridge, where the queen was met by the mayor on the broad green known as Christ's Pieces. From there Anne led her court and a host of trailing officials to Trinity Lodge, home of Richard Bentley, the master of Trinity College. Three aging men were summoned and, on entering the room, they knelt before the queen. A royal proclamation was read and a sword drawn. The blade was touched to each man's shoulders, thus creating three new knights of the realm: Sir John Ellis, Sir James Montague, and Sir Isaac Newton. A sumptuous dinner was later given for the queen in newly redecorated Trinity Hall, where the most illustrious guest after Anne herself had waited tables and taken his meals some forty-five years earlier.

The royal party soon left Cambridge, but Newton lingered for a time. This was a moment to be savored. Shunned by all but a handful of students as a professor, he suddenly found himself the object of their undivided attention. "We always took care on Sunday to place ourselves before him," one of them wrote many years later. "As he sat with the heads of the colleges, we gaz'd on him, never enough satisfy'd, as on someone divine."[3]

Despite his many other duties, Newton had not abandoned his pursuit of science. On February 16, 1704, Secretary Hans Sloane made the following entry in the Journal Book of the Royal Society: "The President presented his book of Optics to the Society. Mr. Halley was desired to peruse it and give an abstract of it to the Society. The Society gave the President their thanks for the Book and for his being pleased to Publish it."[4]

In contrast to the *Principia,* whose publication was preceded by considerable fanfare and much anticipation, nowhere in the minutes of the Royal Society is the *Opticks* mentioned prior to its coming off the press. And it is the only one of Newton's major works that he prepared for publication on his own. Nor did he dedicate this second pillar of his great scientific legacy to the Royal Society, as he had the first.

The reason for his silence becomes clear when one reads the preface. Newton informs the reader that the volume is mainly the product of research completed when he was a young man. "To avoid being engaged in Disputes about these matters, I have hitherto delayed the printing, and should still have delayed it, had not the Importunity of Friends prevailed upon me."[5] To make a long story short, Newton had once again waited for Hooke to pass from the scene before moving forward. His scarcely veiled reference to the deceased curator of experiments was not easily mistaken by those acquainted with the bitter controversy sparked by Newton's stunning paper on light written a generation before.

True to his professional ideals, which had been set out in his "Scheme for establishing the Royal Society," Newton informs the reader of his intentions: "My Design in this Book is not to explain the Properties of Light by Hypothesis, but to propose and prove them by Reason and Experiments."[6] Thus the gifted theoretician, whose genius had shown itself in the *Principia,* laid claim to equal stature as an experimentalist, a rare combination in the history of science.

Whereas the *Principia* is a mathematical work involving intricate geometrical relationships and only a handful of major experiments, the *Opticks* overflows with detailed accounts of reflection and refraction, the separation of white light into the colors of the spectrum, the manner of the eye's operation, the formation of images by lenses, the colors of the rainbow, the construction of the reflecting telescope, and much more. Unable to contain himself, the author introduces many subjects that have little or nothing to do with the behavior and analysis of light: metabolism and digestion, the circulation of the blood, the creation of the world and the Great Flood of Noah, the scientific method, even the images that haunt the dreams of madmen. Moreover, Newton wrote the *Opticks* in English, making its contents accessible to a far wider audience than the *Principia,* whose classical Latin had stymied many potential readers. His friend John Locke, who had only months to live, was baffled by the earlier work but read the *Opticks* "with pleasure, acquainting myself with every thing in them."[7]

Yet even genius has it limits. Newton had delayed publication of the *Opticks* for another reason besides his long and bitter running battle with Robert Hooke. He had always planned to publish the book in four parts rather than the three that finally went to press. In that part of the manuscript he withheld, he wrote of attempting to formulate a single principle with which to account for the behavior of matter: "If Nature be most simple & fully consonant to her self she observes the same method in regulating the motions of smaller bodies [including the corpuscles of light] which she doth in regulating those of the greater [the sun, moon, and planets]."[8] But just as in 1693, when he broke with Fatio de Dullier and saw his alchemical experiments fall apart, he was unable to find the universal principle he was seeking, gravity only being one element of it. The dream of his early manhood had taken the form of a nightmare that had lasted for thirty years. Even the presidency of the Royal Society, the mastership of the mint, a knighthood, and the publication of the greatest books of science ever written could not make it go away.

Eleven

War

Isaac Newton was not the only keeper of a kingdom who arose early and worked deep into the night. John Flamsteed, who, in 1675, was appointed England's first astronomer royal by King Charles II, came calling so early one morning that Newton was still in bed. While waiting for his host to dress, Flamsteed read one of Newton's Bibles and reminded himself that he must keep calm no matter how irritating Newton might be. He considered him to be "a good man at bottom," but highly suspicious and too easily influenced by others. More than anyone else, Flamsteed blamed Edmond Halley for the shabby way Newton had been treating him of late. Looking the part of a dour Puritan minister in the portrait of him that hangs in the Royal Society, Flamsteed suspected that Halley was an atheist, an unbeliever who mocked the religious behind their backs.

Suddenly Flamsteed had what he thought was a brilliant idea. He scribbled a note and hid it in Newton's Bible: "Read Jeremiah, chapter 10 to the 10th verse," a declaration against false prophets and liars. Flamsteed later confided to a friend, "I do not know whether he has seen it, but I think he cannot take it amiss if he has. I have showed him the way of the world much better than politics or a play could do."[1]

While Halley had his jocular moments when he poked fun at the overly zealous, it is doubtful that he rejected organized religion out of hand and more doubtful still that Newton would have associated with him if he had. The real source of Newton's discontent with Flamsteed, which had reached the point where the astronomer no longer attended the meetings of the Royal Society, was Flamsteed's refusal to yield to Newton's every demand. Newton was anxious to publish a second edition of the *Principia,* and he needed data on the motions of the moon and planets that Flamsteed was gathering at the Royal Observatory in Greenwich, a pleasant boat ride down the Thames from London.

Had Newton been more diplomatic, there is little doubt that Flamsteed would have been more cooperative. For all his bluster, and despite his roughshod treatment at Newton's hands, Flamsteed was in awe of the man and simply wanted to be treated with dignity. Moreover, Flamsteed was inching his way toward the completion of a work upon which his own fame would rest, the projected *Historia Coelestis* (History of the Heavens), the most ambitious catalog of the stars ever undertaken.

When Flamsteed had asked Newton to visit Greenwich to see firsthand what he was doing, Newton declined the invitation, as had Halley. Flamsteed then called on Newton at his home, only to be rejected a second time. No shrinking violet, the astronomer royal proceeded to lecture Newton, who was rarely challenged to his face. Comparing himself to those who labored on London's great St. Paul's Cathedral, Flamsteed complained: "I had hewed the materials out of the rock, brought them together, and formed them" and "that hands and time were to be allowed to perfect the building and cover it."[2] Newton grudgingly admitted that he had a point, but once Flamsteed was out of sight the quarrel escalated into war.

Whereas Flamsteed accused Newton of sending his young disciples to Greenwich to spy on his progress, Newton railed against Flamsteed's secrecy and intentional dawdling. What rankled Newton even more was the fact that for the first time in his professional life someone else was in control: Flamsteed possessed something that Newton had wanted desperately for years, and Flamsteed well knew it.

Grasping the nettle, Newton went at last to Greenwich to dine in April 1704. Relishing the circumstances, an unwise Flamsteed informed Newton of some faults he had uncovered in the fourth book of the *Principia,* "which instead of thanking me for, he resented ill." When the conversation then turned to the recently issued *Opticks,* a copy of which Newton had sent to Greenwich, things went no smoother. "I thanked him for his book: he said then he hoped I approved it. I told him truly no." Yet Flamsteed admitted privately that "upon trial all the experiments succeeded as Newton related them, which kept me quiet and in suspense."[3]

When Prince George visited Greenwich a few months later, he offered to finance the printing of Flamsteed's yet unfinished masterpiece. Newton, who was not to be denied, got wind of the proposal and moved swiftly to make certain that the astronomer royal was in no position to refuse. At the next meeting of the Royal Society the president persuaded the Fellows to "encourage" Flamsteed's publication. In another cunning move, Newton saw to it that Prince George was elected a Fellow the following week. Though he sputtered and squawked, Flamsteed was forced to accept the royal offer, for princes were not to be flouted.

Even more galling was the fact that Newton, who had become Flamsteed's open enemy, was appointed head of a commission to examine his papers and determine which of them were fit to print. "With these persons," Flamsteed lamented, "Sir Isaac Newton began to act his part, and carry on his designs."[4] Savoring his triumph, Newton drafted a letter commanding Flamsteed to bring specimens of his work to a dinner with the commissioners, signing himself "Your very loving Friend & humble servant."[5]

The next confrontation occurred when Flamsteed learned that rather than publish all of his papers—including every major star catalog from the Greek astronomer Claudius Ptolemy down to Edmond Halley—Newton wanted to print only that material of concern to him. He was especially interested in Flamsteed's data on the motions of the moon, which were critical to the refinement of his work on gravity. It was at this point that Flamsteed decided to block publication any way he could. Deadlock set in and Newton, try as he might, could not get Flamsteed to budge. Not until April 1708, two years after the project began, did the first page of Volume I of the *Historia Coelestis* emerge from the press. It would be another four years before what Flamsteed considered to be his mutilated work was finally between covers. Meanwhile, the bickering and the sniping continued. Newton, who never truly succeeded in bending Flamsteed to his will, had the astronomer's name stricken from the list of Royal Society Fellows in 1709 for nonpayment of dues. Flamsteed responded by drafting a bitter letter to his friend Abraham Sharp: "Newton has been lately much talked of; but not much to his advantage. Our society is ruined by his close, politic, and cunning forecast."[6]

In truth, Flamsteed was indulging in some wishful thinking. Whatever else Newton might be guilty of, he had almost single-handedly pulled the society back from the brink of disaster while restoring its tarnished reputation. He was turning it into an international enterprise as well. Foreign membership more than doubled during his presidency and included almost as many political figures as natural philosophers. He was especially fond of ambassadors, who, like other distinguished visitors, were seated in comfortable armchairs near the president and entertained by experiments carefully chosen for their spectacular effects. On one such occasion Senior Grimani, the Venetian ambassador, Signor Gerardini, envoy from the Grand Duke of Tuscany, and the Duke D'Armont, the French ambassador, were shown two preparations "made of the veins and arteries of a human liver by injecting red wax into them, which was very curiously and beautifully performed."[7] They also saw light produced by friction and were treated to a demonstration of the power of gravity in a vacuum. But mostly those who served kings and queens and princes came to look upon Newton himself, who had attained the stature of a living god.

After occupying several different residences, Newton finally settled at 35 St. Martin's Street in the parish of Westminster, now a part of greater London. Leicester House, as this stone dwelling became known, was a three-story structure to which its new occupant is believed to have added an observatory. Whether by Catherine Barton's choice or his own, Newton was surrounded with furnishings of red. The chairs and sofa were covered in red upholstery; the draperies were of the same color; red pillows and a matching coverlet decorated his bed, next to which stood a red settee on which he napped after coming home from the mint. Among his other new furnishings were an expensive bell for summoning servants, two bottle racks, three beer stands, and a wooden tub. Of his plans for the tub, a widow living nearby is said to have reported that "every morning he takes his seat in front of a tub of soap-suds, and occupies himself for hours blowing soap-bubbles through a common clay-pipe intently watching till they burst."[8] Like so many stories about the great man, this one seems more fiction than fact, but it is nice to imagine a Newton human enough to delight in this childhood pleasure.

The second edition of the *Principia,* to be followed by a third, was published in 1713, much to the relief of anxious scholars. The first edition had been limited to a few hundred copies and these had long since sold out. For those wealthy enough, a used volume could be had for the considerable sum of two guineas. Those less well off were reduced to the tedious labor of copying it out page by page.

The task of publishing the work so admirably performed by Halley for the first edition now fell to Richard Bentley, a classical scholar and master of Trinity College. Bentley's motives were less pure than Halley's had been, for he made it clear that it was the love of money rather than of natural philosophy that caused him to undertake the work. Moreover, Bentley was not grounded in the finer points of analytical geometry and had to recruit Roger Cotes, a young professor of astronomy and natural philosophy, to deal with the many technical matters. The enterprise was further tainted by Newton's ill treatment of Flamsteed. He had hounded the astronomer royal for years, employing virtually every dirty trick imaginable in an effort to gain control of his lifework. After expelling the astronomer from the Royal Society, he systematically pirated his observations for what he believed to be a "higher" purpose.

Only years later did Flamsteed gain a measure of revenge. He undertook the printing of a complete edition of the *Historia Coelestis,* restoring all the star catalogs and other papers Newton had so shamefully omitted. Then, for good measure, Flamsteed purchased every available copy printed by Newton and cast them into the fire, a ritual act of purification.

Already involved in a long and brutal war on one front, Newton became engaged in another. If judged solely on looks, his latest opponent seemed little more threatening than the bent and twisted Hooke. Indeed, Gottfried Wilhelm von Leibniz was stooped at the waist and of medium height; his broad shoulders sat atop a thick torso propelled about by bandy legs and tiny feet. Beneath his dark eyes and long fleshy nose was a large mouth that turned up at the corners, like the cat who has secretly swallowed the canary. Yet it was not Leibniz's looks that mattered in his coming battle with Newton but his wits, which Leibniz, a genius in his own right, possessed in plenty.

Remembered more for his works as a philosopher than anything else, the tireless German scholar was equally well versed in the fields of theology, history, mathematics, law, mechanics, languages, and geology. While Newton was drafting the first papers on the calculus that would one day lock the two in mortal combat, twenty-year-old Leibniz was composing a brilliant treatise entitled *On the Art of Combination,* a model of logical analysis that has been credited as the ancestor of the modern computer. Equally skilled with his hands, the young German constructed an automatic calculating machine, which he carried to London for the education and pleasure of the Royal Society Fellows, who gladly welcomed him into their company, as they had Newton on the delivery of his telescope.

Leibniz's rapid advance as a mathematician was second only to Newton's. By the waning months of 1675, he had developed the same fluxional method, or calculus, that Newton had formulated a decade earlier. This was not a case of theft, as Newton would later charge, but of independent formulation on Leibniz's part. Moreover, he introduced a superior system of notation in his equations, which eventually replaced Newton's more cumbersome method and is still in use today.

As one of the few who knew of Newton's mathematical genius early on, John Collins had pleaded with him to publish. But Newton preferred secrecy to publicity and missed his chance to establish clear-cut priority. To make matters worse, Leibniz had visited London in 1676. Unbeknown to Newton, the visitor was given access to his mathematical papers—including the calculus—by Collins, who was deeply impressed by the level of the German's scholarship. Collins also wanted Leibniz and his fellow Europeans to know just how great Newton was as a mathematician. Leibniz proceeded to take notes from certain of the papers, but only later would it be discovered that he had copied nothing of the calculus. The simplest explanation for his behavior is also the most convincing: Leibniz had already mastered the method on his own, so Newton had nothing to teach him on the subject. He may even have prided himself on the superiority of his notational system. Only after Collins had gone to his grave years later would Newton learn of what was to him a great betrayal.

Meanwhile, Newton and Leibniz entered into a correspondence on mathematics. By his own admission, Newton was more than a little impressed by Leibniz's grasp of the subject. Still the Englishman was so sure of himself that he refused to discuss the calculus openly, believing that no one else could achieve what he had accomplished as a young man in his early twenties. At the same time, Leibniz, who knew Newton's secret from having visited London, confided to his friend Otto Menke: "Mr. Newton developed it [the calculus], but I arrived at it by another way. One man makes one contribution, another man another."[9]

The inevitable storm broke in full fury in 1699, after Leibniz published two papers on the revolutionary mathematics. He further suggested in print that it was Newton who owed a debt to him, an erroneous claim that was bound to provoke a heated response. Not only had Newton been first, but he had solved two of the most difficult challenge problems Johann Bernoulli could devise, and Leibniz, who was Bernoulli's friend, well knew it.

Though his close ties to Newton had been severed, Fatio de Dullier rushed to his defense. He stated in print that Newton was not only the "first inventor of the calculus," he was the earliest by many years. Whether Leibniz was the second inventor he would leave to the judgment of others. No one who examines the records will be deceived by Leibniz's claim "to the invention of the calculus for himself."[10]

Leibniz, who was deeply angered by the suggestion that he had stolen from Newton, wrote a sharp letter of protest to the Royal Society. Still not satisfied, he penned a review for the *Acta Eruditorum* (Transactions of the Learned), refuting Fatio's charges and laying the blame for the attack on his honor at

Newton's door. No less angered himself, Newton was heard to thunder more than once: "Second inventors count for nothing!"[11]

The conflict entered a new and even more brutal stage following the publication of the *Opticks* in 1704. Two mathematical papers were printed at the end of the work, and in the preface Newton wrote that years ago he had lent out a manuscript of these papers and implied that Leibniz had stolen from it. Leibniz responded by writing a second review for the *Acta* in an attempt to turn the tables. Newton's fluxions, Leibniz charged, were merely the calculus by another name. Like a master counterfeiter, Newton had disguised his forged coin by devising a clever method of notation to conceal his debt to Leibniz.

So the war went, measure for measure, blow for blow, with many of them being struck by the undignified disciples of two geniuses in language that today would result in charges of libel. Finally, a weary Leibniz had had enough and decided to appeal to the ultimate forum, the Royal Society itself.

His long letter was read at the meeting of January 31, 1712, and duly entered in the record. He had endured enough "empty and unjust braying" on the part of Newton's defenders. Even his rival must disapprove of the unending assault on his honor: "I am confident that he will freely give evidence of his opinion on this issue."[12]

Too late did Leibniz realize that he had committed a huge blunder; indeed, he had played directly into Newton's hands. The president hit upon a devilishly clever scheme. Leibniz wanted justice. Very well, let him have it. The Royal Society would appoint a special committee to settle the nagging priority issue—winner take all.

Newton took great pains to make it appear as if the committee was impartial. Eleven men of differing backgrounds, occupations, and political loyalties were chosen to weigh the evidence. He boasted that the group was "skillful and composed of Gentlemen of several nations, and the Society are satisfied in their Fidelity without adding, omitting or altering any thing in favor of either Party."[13] In truth, Newton had handpicked every committee member, stacking the deck against a faraway Leibniz from the beginning.

Newton's treachery was compounded by the fact that the committee's detailed report was ready in a month and a half, although the last member was appointed only a week before its release. The names of those who served appear nowhere in the document, nor did their identities become public knowledge until a century later, when the Royal Society's minutes were published. Even so, none of this mattered in light of the later discovery that it was Newton himself who compiled the evidence and drafted the report that was issued under the title *Commercium Epistolicum* (Letter on Trade). Every move Leibniz had made, from his London visits to his choice of notation for the calculus, was cast in a suspicious light.

Aside from the relatively few who knew better, it appeared that Leibniz was an intellectual thief who had cleverly stolen Newton's creation and claimed it as his own. The Fellows approved the report to a man and agreed that it should be published as soon as possible. Nine months later, in January 1713, the *Commercium* came off the press. Newton personally saw to it that copies reached the institutions and persons that mattered most, especially Leibniz's European colleagues, who looked upon him as a saint among mathematicians.

Leibniz died three years later, and it would take more than two centuries before the whole truth behind this ugly story became known. Even after his enemy was no more, Newton could not let go. A friend of William Whiston, Newton's successor as Lucasian professor, reported that he once heard Newton pleasantly remark that "He had broke Leibniz's Heart with his Reply to him."[14]

Twelve

"Like a Boy on the Seashore"

JOHN CONDUITT, THE SON OF A WEALTHY FAMILY FROM THE COUNTY of Hampshire in southern England, became a familiar face at Leicester House in the summer of 1717. While there is no doubt that the young man was in awe of Sir Isaac Newton, the primary reason for his frequent visits had nothing to do with science, mathematics, or the operations of the mint. Conduitt came to 35 St. Martin's Street to court Catherine Barton, who at thirty-eight years of age was nine years older than her suitor, but still very beautiful.

Conduitt had served as a captain with the British Army in Spain, where he identified the site of Carteia, a lost city once occupied by the Romans. When word of his discovery reached the Royal Society back in London, he was invited to read a paper on the subject, which he did upon his return to England three months later. It was there that he first met Newton, who may have invited him to Leicester House for dinner at which Catherine would have presided as hostess.

The Hampshire gentleman and the independent-minded beauty were wed on August 26, after a courtship lasting only a few weeks. Catherine bore the couple's first and only child, a daughter, two years later. They baptized her Catherine and nicknamed her Kitty to avoid confusion with her mother. Tradition holds that the Conduitts lived with Newton, who doted on his grandniece. However, Catherine's letters suggest that much of their time was spent at Cranburg Park, the ancestral home of her devoted husband. After all, her uncle was protective of his privacy, which would have been disturbed with a noisy toddler in his midst. The family paid frequent visits during which Conduitt asked Sir Isaac questions for a biography he planned to write of him, the extensive notes for which survive. Newton formed a high opinion of his nephew-in-law and began laying plans for the day when Conduitt would succeed him as master of the mint.

In 1722, as Newton was nearing his eightieth birthday, he received a painful reminder of his mortality. Taken seriously ill with kidney stones, he was gradually nursed back to health by his personal physician and Royal Society Fellow, Dr. Richard Mead. "I am getting well little by little," he wrote a friend, but in truth he had entered a period of irreversible decline.[1] There is little to indicate that he had mellowed with age. It was not his way. He still relished a quarrel and fought on in print with the followers of Leibniz, although the other creator of the calculus was long in his grave. He now experienced the ultimate triumph of outliving those who had dared to challenge or cross him. He was internationally famous, and his hard-won glory made it worthwhile to endure.

There were many requests for audiences with the old giant, but Catherine, who had moved back to London with her husband and growing daughter, kept most of the curious at bay.

Among them was a young man from the American colonies named Benjamin Franklin, who was just mastering the printer's trade. Franklin wrote in his diary of being promised "an opportunity of some time or other of seeing Sir Isaac Newton, of which I was extremely desirous; but this never happened."[2] Years later, after he became famous for his brilliant experiments on electricity, Franklin was awarded the Copley Medal, the highest prize given by the Royal Society.

Another of those disappointed at not meeting Newton was a young Frenchman named François Marie Arouet, who wrote under the pen name Voltaire. The closest the critic of royal power in France got was a meeting with Catherine. She matched wits with her sharp-tongued visitor, whose ears pricked up when she told him stories of her uncle and his sometimes peculiar habits. Voltaire, who had the morals of a gossip columnist, repaid her kindness by spreading scandal about his hostess in the drawing rooms of London and Paris, not to mention in his books.

Among the few allowed to visit Newton was William Stukeley, the young doctor from Lincolnshire. He credited Newton's long life to a strong constitution and regular habits. In 1725 Stukeley watched in amazement as Newton added up a column of figures without the aid of spectacles, pen, or ink. He told Stukeley that his breakfast consisted of boiled orange peel, sweetened tea, and a little bread with butter. He drank more water than when he was a young man and took only a small amount of wine with his dinner. His other meals consisted mostly of broth, vegetables, and fruit, "of which he always ate very heartily."[3] Never a large meat eater, Newton had virtually given it up in recent years. Still plagued by kidney stones, he also gave up riding in his carriage, whose jostling over London's cobblestone streets caused him extreme discomfort. He was carried about by four men in a slower but smooth-riding sedan chair, his arms dangling out of the openings on either side. His eyes remained lively and piercing, while his silver hair, which seemed never to have thinned, was described by Stukeley as a "venerable sight" when his wig was removed. More remarkable still for one who loved sweets and grew plump from eating them, Newton had lost only a single tooth.

After Catherine's marriage, Newton lived much as he always had, taking his meals alone and passing most of his days in his study, reading and writing. His once heavy duties at the mint were lightened by Conduitt, and for a year before his death he rarely visited the tower. His excellent record of attendance at the Royal Society became tarnished as age took its toll. More and more the mace remained in its stand while the vice-president occupied the chair.

On occasion Newton surprised those who knew him by turning up at unexpected places. One evening a gathering of men from Lincolnshire was held at the Ship Tavern. Stukeley was in the upstairs dining room where "the better sort of company" had met when someone reported that an old gentleman seen downstairs was thought to be Sir Isaac Newton. The disbelieving doctor rushed down and, to his complete surprise, spied Newton seated alone. Word was sent up to the second floor, which soon emptied. Surrounded by a captive audience, Newton told stories and gave his opinion on many subjects, including opera, which was just becoming popular. "'I went to the last opera,' says he, 'The first act gave me pleasure, the second quite tired me: at the third I ran away.'"[4] His opinion of poetry was no higher; he called it "a kind of ingenious nonsense."[5]

Three days later, Stukeley breakfasted with Newton and Halley, who had replaced John Flamsteed as astronomer royal after

Flamsteed's death. Newton used the occasion to blast the fallen astronomer yet again. Flamsteed, he fumed, had provided him with only a handful of observations when he was struggling to complete his lunar theory. He owed the man nothing for whatever success he had enjoyed on that score. He boasted that he could now finish his work on the moon if he wanted to, "but rather he chose to leave it to others." On another occasion he told Halley he was thinking of having one more "shake at the moon." And to Benjamin Smith, his nephew, he once spoke of having yet "another touch at metals."[6] Such talk was based on the wistful thinking of an old man to which Newton, of all persons, was certainly entitled. Yet these fancies are revealing for another reason. Newton never believed that his natural philosophy had been brought to a satisfactory conclusion, a painful thought he lived with to the end.

Surprising as it may seem for a man of his careful habits, Newton drafted no will. Besides having given to various charities and to the Royal Society, much of his fortune had already been expended on his relatives, many of whom were extremely careless with money. His nieces and nephews by his stepbrother and stepsisters did very well, as did the Conduitts and their daughter Kitty, who received an estate in Kensington from her great-uncle valued at 4,000 pounds.

On the last day of February 1727, Newton came into London to preside at the March 2 meeting of the Royal Society. Conduitt had not seen him looking so fit in years and told him so. A smiling Newton replied that he had slept the Sunday before from eleven at night to eight in the morning without waking. But by the time he returned home it was apparent that the rigors of travel had been too taxing. Another stone lodged itself in his bladder, and the two doctors called in by Conduitt held out no hope for recovery. Extreme pain alternated with brief periods of tranquility during the next several days. "Though the drops of sweat ran down his face," an anguished Conduitt wrote, "he never complained, or cried out, or shewed the least signs of peevishness or impatience."[7]

Stukeley painted an even more dramatic picture of the death scene: The pain "rose to such a height that the bed under him, and the very room, shook with his agonys to the wonder of those that were present. Such a struggle had his great soul to quit its earthly tabernacle!"[8] It was clearly with his soul that Newton remained most concerned. He refused the final rites of the church, the last act of a man who, for more than half a century, secretly viewed the concept of the Trinity with horror. To yield at the last minute would give Satan the triumph so long denied him by his unbending foe.

On Wednesday, March 15, Newton rallied, raising hopes that he might prove the experts wrong once again. But a few days later he slipped into a coma, remaining unconscious until he died between one and two o'clock in the morning of March 20, at age eighty-four. Not long before his passing he remarked, "I do not know what I may appear to the world; but to myself I seem to have been only like a boy, playing on the seashore, and diverting myself in now and then finding a smoother pebble or prettier shell than ordinary, while the great ocean of truth lay all undiscovered before me."[9] Strangely, Albert Einstein, the greatest scientist since Newton, also pictured himself as a child asking childlike questions: "What would the world look like if I could ride on a beam of light?" It was Einstein who wrote of Sir Isaac, "Fortunate Newton, happy child of science. Nature to him was an open book, whose letters he could read without effort."[10]

On March 23 the following entry, simple yet touching, appeared in the Journal Book of the Royal Society: "The Chair being Vacant by the Death of Sir Isaac Newton there is no Meeting this Day."[11] Newton's body lay in state in Westminster Abbey until April 4, the day of the funeral. The coffin was carried into the main sanctuary by the Lord High Chancellor, the Dukes of Montrose and Roxborough, and the Earls of Pembroke, Sussex, and Macclesfield, all noblemen and all members of the Royal Society. They were followed by the procession of mourners led by Sir Michael Newton, a knight and distant relative of the fallen giant. The service was performed by the bishop of Rochester.

Newton's final resting place was chosen for its prominence in the nave of the great abbey. Nearby lie the stilled bards of Poet's Corner—Chaucer, Robert Browning, and Alfred Lord Tennyson. Closest to Newton is the great astronomer Sir John Herschel, who opened the night sky with telescopes as never before. Time, like a slow and ancient glacier, washes through the sanctuary, its flow wearing the pavement stones into small hollows. Its work is aided by the countless thousands of awed visitors who come each year to pause in silent tribute to the sacred permanence of the dead.

Notes

ABBREVIATIONS OF FREQUENTLY USED SOURCES

I.N. *Corres.*	*The Correspondence of Isaac Newton.* Ed. by H. W. Turnbull, J. F. Scott, A. R. Hall, and Laura Tilling. 7 vols. Cambridge: Cambridge University Press, 1959–77.
Keynes MS.	Keynes Manuscript Collection, King's College Library, Cambridge, England.
U.L.C. Add. MS.	University Library Cambridge, Additional Manuscript, Cambridge, England.

CHAPTER ONE: "TO PLAY PHILOSOPHICALLY"

1. Michael B. Young, *Charles I* (New York, 1997), 172.
2. Canon C. W. Foster, "Sir Isaac Newton's Family," *Reports and Papers of the Architectural Societies of the County of Lincoln, County of York, Archdeaconries of Northampton and Oakham, and County of Leicester,* XXXIX, pt. I (1928), 45–47.
3. Colsterworth Village Church: Parish Register (January 1, 1643).
4. Richard S. Westfall, "Short-Writing and the State of Newton's Conscience, 1662 (I)," *Notes and Records of the Royal Society of London*, 18 (June 1963), 13.
5. Keynes MS. 136, 4.
6. Ibid.
7. Ibid., 5.
8. U.L.C. Add. MS. 3975, 178.
9. Keynes MS. 136, 6–7.
10. Ibid., 6.
11. Ibid.
12. Ibid.

CHAPTER TWO: "MY GREATER FRIEND"

1. Thomas Fuller, *History of the University of Cambridge published with his Church-history of Britain* (London, 1655), 122.
2. U.L.C. Add. MS. 3996. For a modern reproduction of the notebook see *Certain Philosophical Questions: Newton's Trinity Notebook*, ed. by J. E. McGuire and T. Martin (Cambridge, Mass., 1983).
3. Ibid.
4. Ibid., ff. 88–89.
5. Ibid., f. 101^{v}.
6. Keynes MS. 130 (10).
7. Roger North, *The Lives of the Norths*, ed. by Augustus Jessopp, 3 vols. (London, 1890), III, 284.
8. I.N. *Corres.*, III, 153.
9. U.L.C. Add. MS. 3996, f. 128.
10. U.L.C. Add. MS. 4007, ff. 706–7.
11. Keynes MS. 130 (7).
12. Keynes MS. 137.
13. Trinity College, Cambridge, MS. R4.48^{c}.
14. Keynes MS. 136 (8).

CHAPTER THREE: OF GENIUS, FIRE, AND PLAGUE

1. Thomas Vincent, *God's Terrible Voice in the City* (London, 1667), 29.
2. U.L.C. Add. MS. 3968.41, f. 85^{r}.
3. William Whiston, *Memoirs of the Life and Writings of Mr. William Whiston*, 2 vols. (London, 1749), I, 39.
4. William Stukeley, *Memoirs of Sir Isaac Newton's Life*, ed. by A. Hastings White (London, 1936), 19–20.
5. Henry Pemberton, *A View of Sir Isaac Newton's Philosophy* (London, 1728), Preface.
6. U.L.C. Add. MS. 3968.41, f. 85^{r}.
7. Whiston, 36–37.
8. I.N. *Corres.*, I, 92.
9. U.L.C. Add. MS. 3975, 12.
10. I.N. *Corres.*, I, 92.
11. Canon C. W. Foster, "Sir Isaac Newton's Family," *Reports and Papers of the Architectural Societies of the County of Lincoln, County of York, Archdeaconries of Northampton and Oakham, and County of Leister*, XXXIX, pt. I (1928), 3.

CHAPTER FOUR: THE REVOLUTIONARY PROFESSOR

1. Fitzwilliam Museum Notebook, Fitzwilliam Museum, Cambridge, England, n.p.
2. Keynes MS. 130 (10), f. 2^{v}.
3. I.N. *Corres.,* I, 13.
4. Ibid., 14–15.
5. U.L.C. Add. MS. 4002, 1.
6. Quoted in *Correspondence of Sir Isaac Newton and Professor Cotes*, ed. by J. Edleston (London, 1950), xlii.
7. U.L.C. Add. MS. 4002, 22.
8. I.N. *Corres.*, I, 3–4.
9. Ibid., 89.
10. Ibid., 82–83.
11. Ibid., 84.
12. Ibid., 102.
13. Ibid., 107.
14. Ibid., 109.

CHAPTER FIVE: "KINDLING COAL"

1. *The Diary of Samuel Pepys*, ed. by Robert Latham and William Matthews, 11 vols. (Berkeley and Los Angeles, 1972), VI, 36–37.
2. John Aubrey, *Brief Lives*, ed. by O. L. Dick, 3rd ed. (London, 1968), 165.
3. I.N. *Corres.*, I, 100.
4. Ibid., 110–11.
5. Ibid., 144.
6. Ibid., 109.
7. Ibid., 151.
8. Ibid., 171–72.
9. Ibid., 262.
10. Ibid., 282.
11. Ibid., 294–95.
12. Ibid., 328–29.
13. Ibid., 366.
14. *The Diary of Robert Hooke, 1672–1680*, ed. by Henry W. Robinson and Walter Adams (London, 1968), 200, 205–6.
15. I.N. *Corres.*, I, 405.
16. Ibid., 412.
17. Ibid., 413.
18. Ibid., 416.
19. Ibid., II, 6.

CHAPTER SIX: THE ALCHEMIST

1. Keynes MS. 137.
2. Keynes MS. 135.
3. Ibid.
4. Gale E. Christianson, *In the Presence of the Creator: Isaac Newton and His Times* (New York, 1984), 204.
5. Isaiah, 45:3.
6. Keynes MS. 33, f. 5^{v}.
7. Richard S. Westfall, "Short-Writing and the State of Newton's Conscience, 1662 (I)," *Notes and Records of the Royal Society of London* (June 1963), 13.
8. Yahuda Manuscript Collection, Jewish National and University Library, Jerusalem, Israel, MS. 1.1, f. 1.
9. Ibid., f. 16^{r}. Newton quoting from Revelations 21:8.

CHAPTER SEVEN: "A BOOK NOBODY UNDERSTANDS"

1. I.N. *Corres.*, II, 433–35.
2. Ibid.
3. Keynes MS. 135.
4. Ibid.
5. Ibid.
6. Ibid.
7. U.L.C. Add. MS. 4004, ff. 707^{r}–707^{v}.
8. Ibid.
9. I.N. *Corres.*, II, 413.
10. Quoted in Richard S. Westfall, *Never at Rest: A Biography of Isaac Newton* (Cambridge, 1980), 452.
11. I.N. *Corres.*, II, 435–37.
12. Keynes MS. 130 (6), Book 2.
13. Isaac Newton, *Mathematical Principles of Natural Philosophy*, trans. and ed. by Andrew Motte, 1729. Revised by Florian Cajori (Berkeley and Los Angeles, 1934), 415. See the Further Reading section in this book for a more recent translation by I. Bernard Cohen and Anne Whitman.
14. Ibid., 435.

CHAPTER EIGHT: "YOUR MOST UNFORTUNATE SERVANT"

1. I.N. *Corres.*, II, 483.
2. Keynes MS. 130 (5).
3. Ibid., 135.
4. I.N. *Corres.*, III, 279.

5. Ibid., 280.
6. Ibid., 283.
7. Ibid., 283–84.
8. Louis Trenchard More, *Isaac Newton: A Biography* (London, 1934), 389.

CHAPTER NINE: MARK OF THE LION

1. I.N. *Corres.*, IV, 188.
2. *Guzman Redivivus. A Short View of the Life of Will Chaloner* (London, 1699), 4.
3. I.N. *Corres.*, IV, 307.
4. John Craig, "Isaac Newton and the Counterfeiters," *Notes and Records of the Royal Society of London*, 18 (1963), 139.
5. *The Mathematical Papers of Isaac Newton*, ed. by D. T. Whiteside, 8 vols. (Cambridge, 1981), VIII, 9.
6. *The Diary of John Evelyn*, ed. by E. S. De Beer, 6 vols. (Oxford, 1955), V, 284, fn. 5.
7. I.N. *Corres.*, IV, 339.
8. Jonathan Swift, *Journal to Stella*, ed. by Harold Williams (Oxford, 1948), 229–30.
9. Brooke Taylor, *Contemplatio philosophica* (London, 1793), 93.
10. I.N. *Corres.*, IV, 382–83.

CHAPTER TEN: THE ROYAL SOCIETY

1. U.L.C. Add. MS. 4005.2, f. 1.
2. William Stukeley, *Memoirs of Sir Isaac Newton's Life*, ed. by A. Hastings White (London, 1936), 78.
3. Ibid., 10.
4. Journal Book (Copy) of the Royal Society, London, England, X, 63–64.
5. Isaac Newton, *Opticks Or A Treatise of the Reflections, Refractions, Inflections and Colours of Light* (New York, 1952), ccxi. See the Further Reading section in this book for a more recent edition.
6. Ibid., 1.
7. Jean-Theophile Desaguliers, *Course of Experimental Philosophy*, 3rd ed. (London, 1763), viii.
8. U.L.C. Add. MS. 3970.3, f. 336.

CHAPTER ELEVEN: WAR

1. Francis Bailey, *An Account of the Rev[d] John Flamsteed, the First Astronomer Royal* (London, 1835), 174–75.
2. Ibid.

3. Ibid., 72, 216.
4. Ibid., 77.
5. I.N. *Corres.*, IV, 432.
6. Bailey, 272.
7. Charles R. Weld, *A History of the Royal Society*, 2 vols. (London, 1848), I, 419, and Journal Book (Copy) of the Royal Society, London, England, X, 483.
8. H.D. Anthony, *Sir Isaac Newton* (London, 1966), 197.
9. Gottfried Wilhelm Leibniz, *Sämtliche Schriften und Briefe*, ser. I (Berlin, 1923), IV, 475–77.
10. Nicolas Fatio de Dullier, *Lineae brevissimi descensus investigatio geometrica duplex* (London, 1699), 18.
11. U.L.C. Add. MS. 3968.37, f. 539ʳ.
12. I.N. *Corres.*, V, 207–8.
13. Journal Book (Copy) of the Royal Society, X, 369, 375, 377, 386.
14. William Whiston, *Historical Memoirs of the Life of Dr. Samuel Clarke* (London, 1730), 132.

CHAPTER TWELVE: "LIKE A BOY ON THE SEASHORE"

1. I.N. *Corres.*, VII, 205.
2. *The Autobiography of Benjamin Franklin*, ed. by Leonard W. Labaree, Ralph L. Ketcham, Helen C. Boatfield, and Helene H. Fineman (New Haven, 1964), 97.
3. *Collections for the History of the Town and Soke of Grantham*, ed. by Edmund Turnor (London, 1806), 165.
4. William Stukeley, *Memoirs of Sir Isaac Newton's Life*, ed. by A. Hastings White (London, 1936), 13–14, 66.
5. Richard de Villamil, *Newton: The Man* (London, 1931), 10.
6. Keynes MS. 130 (6).
7. Ibid., Book I.
8. Turnor, 166; Stukeley, 82–83.
9. Turnor, 173, fn. 2.
10. Isaac Newton, "*Opticks Or A Treatise of the Reflections, Refractions, Inflections, and Colours of Light* (New York, 1952), Foreward, lix.
11. Journal Book (Copy) of the Royal Society, London, England, XIII, 62.

Further Reading

Note: The literature on Newton is a vast and rapidly expanding one. While the following bibliography is far from comprehensive, it should address the curiosity of readers seeking knowledge beyond that imparted in these pages.

Andrade, Edward Neville da Costa. *Sir Isaac Newton*. London: Collins, 1954.

Anthony, H. D. *Sir Isaac Newton*. London: Abelard-Schuman, 1960.

Ariew, Roger. *The Correspondence of G. W. Leibniz and Samuel Clarke*. Indianapolis: Hackett, 2000.

Bailey, Francis. *An Account of the Rev[d] John Flamsteed, the First Astronomer Royal*. London, 1835.

Bate, John. *The Mysteries of Nature and Art*. 3rd ed. London: Andrew Crooke, 1654.

Berlinski, David. *How Newton Unlocked the System of the World*. New York: The Free Press, 2000.

Bixby, William. *The Universe of Galileo and Newton*. New York: American Heritage Publishing Company, 1964.

Boorstin, Daniel J. *The Discoverers*. New York: Random House, 1983.

Brewster, David. *Memoirs of the Life, Writings, and Discoveries of Sir Isaac Newton*. New York: Johnson Reprint Corporation, 1965.

Butts, Robert E., and John W. Davis. *The Methodological Heritage of Newton*. Toronto: University of Toronto Press, 1970.

Challis, C. E., ed. *A New History of the Royal Mint*. Cambridge: Cambridge University Press, 1992.

Christianson, Gale E. *In the Presence of the Creator: Isaac Newton and His Times*. New York: Free Press, 1984.

———. "Newton, the Man—Again." In *Newton's Scientific and Philosophical Legacy*, eds. P. B. Scheuer and G. Debrock. Dordrecht: Kluwer Academic Publishers, 1988: 2–21.

Clark, David H., and Stephen P. H. Clark. *Newton's Tyranny: Suppressed Scientific Discoveries of Stephen Gray and John Flamsteed*. New York: W. H. Freeman, 2000.

Cohen, I. Bernard. *Introduction to Newton's* Principia. Cambridge: Cambridge University Press, 1971.

———, ed. *Isaac Newton's Papers and Letters on Natural Philosophy and Related Documents*. Cambridge, Mass.: Harvard University Press, 1978.

———, and George E. Smith. *Cambridge Companion to Newton*. Cambridge: Cambridge University Press, 2002.

———, and R. S. Westfall. *Newton: Texts, Backgrounds, Commentaries*. New York: Norton, 1995.

Cook, Alan. *Edmond Halley: Charting the Heavens and the Seas*. Oxford: Oxford University Press, 1998.

Craig, John. *Newton at the Mint*. Cambridge: Cambridge University Press, 1946.

De Beer, E. S., ed. *The Diary of John Evelyn*. Oxford University Press, 1955; reprinted 2000.

De Morgan, Augustus. *Newton: His Friend: and His Niece*. London: Dawsons, reprinted 1968.

Desaguliers, Jean-Theophile. *Course of Experimental Philosophy*. London: 1763.

Dobbs, Betty Jo Teeter. *The Foundations of Newton's Alchemy*. Cambridge: Cambridge University Press, 1975.

———. *The Janus Face of Genius: The Role of Alchemy in Newton's Thought*. Cambridge: Cambridge University Press, 1992.

———, and Margaret C. Jacob. *Newton and the Culture of Newtonianism*. Atlantic Highlands, N.J.: Humanities Press, 1995.

Fatio de Dullier, Nicolas. *Lineae brevissimi descensus investigatio geometrica duplex*. London, 1699.

Fauvel, John, et al. *Let Newton Be!* New York: Oxford University Press, 1988.

Feingold, Mordechai, ed. *Before Newton: The Life and Times of Isaac Barrow*. Cambridge: Cambridge University Press, 1990.

Gill, Mary Louise, and James G. Lennox. *Self-Motion from Aristotle to Newton*. Princeton: Princeton University Press, 1994.

Gjertsen, Derek. *The Newton Handbook*. London: Routledge and Kegan Paul, 1986.

Gleick, James. *Isaac Newton*. New York: Pantheon, 2003.

Greenstreet, W. J., ed. *Isaac Newton 1642–1727: A Memorial Volume*. London: G. Bell and Sons, 1927.

Hall, A. R. *All Was Light: An Introduction to Newton's* Optics. Oxford: The Clarendon Press, 1995.

———. *Isaac Newton: Adventurer in Thought*. Cambridge, Mass.: Blackwell, 1992.

———. *Philosophers at War: The Quarrel Between Newton and Leibniz*. Cambridge: Cambridge University Press, 1980.

Hall, Marie Boas. *Henry Oldenburg: Shaping the Royal Society*. Oxford: Oxford University Press, 2002.

Harrison, John R. *The Library of Isaac Newton*. Cambridge: Cambridge University Press, 1978.

Hawking, Stephen, ed. *On the Shoulders of Giants:* Principia. Philadelphia: Running Press, 2005.

Herivel, John. *The Background to Newton's* Principia: *A Study of Newton's Dynamical Researches in the Years 1664–1684*. Oxford: Oxford University Press, 1965.

Hooke, Robert. *An Attempt to Prove the Motion of the Earth from Observations*. London: Royal Society, 1674.

———. *Micrographia: or some Physiological Descriptions of Minute Bodies Made by Magnifying Glasses with Observations and Inquiries thereupon*. London: J. Martin and J. Allestray: 1665.

Hurlbutt III, Robert H. *Hume, Newton, and the Design Argument*. Lincoln: University of Nebraska Press, 1985.

Jacob, Margaret. *The Newtonians and the English Revolution, 1689–1720*. Ithaca, New York: Cornell University Press, 1976.

Jardine, Lisa. *The Curious Life of Robert Hooke: The Man Who Measured London*. New York: HarperCollins, 2004.

Keynes, John Maynard. "Newton the Man." *The Royal Society of London: Newton Tercentenary Celebrations*. Cambridge: Cambridge University Press, 1947: 27–34.

Klawans, Harold L. *Newton's Madness*. New York: Harper and Row, 1990.

Koyré, Alexandre. *Newtonian Studies*. Cambridge, Mass.: Harvard University Press, 1965.

Labaree, Leonard W., et al., eds. *The Autobiography of Benjamin Franklin*. New Haven: Yale University Press, 1964.

Latham, Robert, and William Matthews, eds. *The Diary of Samuel Pepys*. Berkeley and Los Angeles: University of California Press, 1970–1983.

Lerner, Aaron Bunsen. *Einstein and Newton: A Comparison of the Two Greatest Scientists*. Minneapolis: Lerner, 1973.

Manuel, Frank. *A Portrait of Isaac Newton*. Cambridge, Mass.: Harvard University Press, 1968.

———. *Isaac Newton, Historian*. Cambridge, Mass.: Cambridge University Press, 1963.

———. *The Religion of Isaac Newton*. Oxford: The Clarendon Press, 1974.

Maury, Jean-Pierre. *Newton: The Father of Modern Astronomy*. New York: Harry N. Abrams, 1992.

McGuire, J. E., and T. Martin, eds. *Certain Philosophical Questions: Newton's Trinity Notebook*. Cambridge: Cambridge University Press, 1983.

McLachlan, Herbert. *The Religious Opinions of Milton, Locke, and Newton*. New York: Russell and Russell, 1972.

———, ed. *Sir Isaac Newton's Theological Manuscripts*. Liverpool: Liverpool University Press, 1950.

McMullin, Ernan. *Newton on Matter and Activity*. Notre Dame: University of Notre Dame Press, 1978.

Meli, Domenico Bertolini. *Equivalence and Priority: Newton Versus Leibniz*. Oxford: The Clarendon Press, 1993.

Merton, Robert K. *On the Shoulders of Giants: A Shandean Postscript*. Chicago: University of Chicago Press, 1993.

Moore, Louis Trenchard. *Isaac Newton: A Biography*. New York: Scribners, 1934.

Newton, Isaac. *Observations Upon the Prophecies of Daniel and the Apocalypse of St. John*. Whitefish, Mont.: Kessinger, 2003.

———. *Optics, or Treatise on the Reflections, Inflections and Colours of Light*. Buffalo, N.Y.: Prometheus, 2003.

———. *Principia: Mathematical Principles of Natural Philosophy*. Trans. Cohen, I. Bernard and Anne Whitman. Berkeley: University of California Press, 1999.

———. *The Chronology of Ancient Kingdoms Amended*. Whitefish, Mont.: Kessinger, 2004.

Nicolson, Marjorie Hope. *Newton Demands the Muse: Newton's* Optics *and the Eighteenth Century Poets*. Princeton: Princeton University Press, 1946.

North, John David. *Isaac Newton*. Oxford: Oxford University Press, 1967.

Palter, Robert. *The Annus Mirabilis of Sir Isaac Newton, 1666–1966*. Cambridge, Mass.: MIT Press, 1970.

Pemberton, Henry. *A View of Sir Isaac Newton's Philosophy*. London, 1728.

Rattansi, Piyo M. *Isaac Newton and Gravity*. London: Wildwood, 1974.

Richardson, Robert S. *The Star Lovers*. New York: Macmillan, 1967.

Robinson, Henry W. and Walter Adams. *The Diary of Robert Hooke MA MD FRS 1672–1680*. London: Taylor and Francis, 1968.

Scootin, Harry. *Isaac Newton*. New York: Messner, 1955.

Shapin, Steven. *The Scientific Revolution*. Chicago: Cambridge University Press, 1984.

Shapiro, Alan E. *Fits, Passions, and Paroxysms: Physics, Method, and Chemistry and Newton's Theory of Coloured Bodies and Fits of Easy Reflection*. Cambridge: Cambridge University Press, 1993.

———. *The Optical Papers of Isaac Newton.* Cambridge: Cambridge University Press, 1984.

Stukeley, William A. *Memoirs of Sir Isaac Newton's Life.* London: Taylor and Francis, 1936.

Sullivan, J. W. N. *Sir Isaac Newton, 1642–1727.* London: Macmillan, 1938.

Taylor, Brooke. *Contemplatio philosophica.* London, 1793.

Thrower, Norman J. W. ed. *Standing on the Shoulders of Giants: A Longer View of Newton and Halley.* Berkeley: University of California Press, 1990.

Turnbull, H. W., et al. eds. *The Correspondence of Isaac Newton.* Cambridge: Cambridge University Press, 1959–77.

Turnor, Edmund, ed. *Collections for the History of the Town and Soke of Grantham.* London, 1806.

Villamil, Richard de. *Newton: The Man.* New York: Johnson Reprint Corporation, 1972.

Wallis, Peter and Ruth. *Newton and Newtoniana.* Folkestone, England: Dawson, 1977.

Weld, Charles R. *A History of the Royal Society.* London, 1848.

Westfall, Richard S. *Never at Rest: A Biography of Isaac Newton.* Cambridge: Cambridge University Press, 1980.

———. *The Life of Isaac Newton.* Cambridge: Cambridge University Press, 1993.

Whiston, William. *Historical Memoirs of the Life of Dr. Samuel Clarke.* London, 1730.

———. *Memoirs of the Life and Writings of Mr. William Whiston.* London, 1749.

White, Michael. *Isaac Newton: The Last Sorcerer.* Reading, Mass.: Addison-Wesley, 1997.

Whiteside, D. T., ed. *The Mathematical Papers of Isaac Newton.* Cambridge: Cambridge University Press, 1967–80.

Index

5-2-06